ROUTLEDGE LIBRARY EDITIONS:
ETHICS

Volume 31

AN INTRODUCTION TO ETHICS

AN INTRODUCTION TO ETHICS

J. D. MABBOTT

Taylor & Francis Group

LONDON AND NEW YORK

First published in 1966 by Hutchinson & Co (Publishers) Ltd

This edition first published in 2021
by Routledge
2 Park Square, Milton Park, Abingdon, Oxon OX14 4RN

and by Routledge
52 Vanderbilt Avenue, New York, NY 10017

Routledge is an imprint of the Taylor & Francis Group, an informa business

© 1966 J. D. Mabbott

All rights reserved. No part of this book may be reprinted or reproduced or utilised in any form or by any electronic, mechanical, or other means, now known or hereafter invented, including photocopying and recording, or in any information storage or retrieval system, without permission in writing from the publishers.

Trademark notice: Product or corporate names may be trademarks or registered trademarks, and are used only for identification and explanation without intent to infringe.

British Library Cataloguing in Publication Data
A catalogue record for this book is available from the British Library

ISBN: 978-0-367-85624-3 (Set)
ISBN: 978-1-00-305260-9 (Set) (ebk)
ISBN: 978-0-367-50914-9 (Volume 31) (hbk)
ISBN: 978-1-00-305182-4 (Volume 31) (ebk)

Publisher's Note
The publisher has gone to great lengths to ensure the quality of this reprint but points out that some imperfections in the original copies may be apparent.

Disclaimer
The publisher has made every effort to trace copyright holders and would welcome correspondence from those they have been unable to trace.

AN INTRODUCTION TO ETHICS

J. D. Mabbott

HUTCHINSON UNIVERSITY LIBRARY
LONDON

HUTCHINSON & CO (Publishers) LTD
3 Fitzroy Square, London W1

London Melbourne Sydney Auckland
Wellington Johannesburg Cape Town
and agencies throughout the world

First published 1966
Reprinted 1967, 1969, and 1974

© J. D. Mabbott 1966

The paperback edition of this book is sold subject to the condition that it shall not, by way of trade or otherwise, be lent, re-sold, hired out or otherwise circulated without the publisher's prior consent in any form of binding or cover other than that in which it is published and without a similar condition including this condition being imposed on the subsequent purchaser.

Printed in Great Britain by The Anchor Press Ltd
and bound by Wm. Brendon & Son Ltd
both of Tiptree, Essex

ISBN 0 09 078850 8 (cased)
     0 09 078851 6 (paper)

ACKNOWLEDGMENT

The author is grateful to the Bodley Head Ltd for permission to quote from *Beasts and Superbeasts* by Saki.

CONTENTS

	Preface	9
1	Introduction	13
2	Pleasure Utilitarianism	15
3	Ideal Utilitarianism	22
4	Moral rules	32
5	Right, ought, and duty	51
6	The duty to think	65
7	Variations in moral belief	70
8	The Emotive theory	74
9	Subjectivism and Objectivism	92
10	Freedom of choice	109
11	Punishment	127
12	Rights and duties	139
13	Conclusion	152
	Index	155

PREFACE

This book is intended as an Introduction in a double sense. It is written not only for readers who have not previously encountered moral philosophy but specifically for those who have read no philosophy at all and who wish to start philosophy with ethics.

There are strong arguments against such a start. The language and ideas of morals present great difficulties; the delimitations of the subject, especially from philosophical psychology, are not easy to determine. Logically it would seem better to start with the analysis of more straightforward factual propositions, together with some study of logic. In my experience, able students starting a full-scale honours course find this latter approach entirely satisfactory. Like the puppies in Plato, they enjoy tearing to bits the ideas thrown to them. But there are many who find this approach barren, negative, destructive, and unsatisfying. They include mature students, amateurs who have no continuous instruction to help them, and undergraduates who are not naturally attracted by or equipped for abstract speculation or a purely linguistic approach. Many of them have also got the idea that philosophers should concern themselves with problems which would seem genuine problems even to a non-philosopher. The problems which arise or are put before them in the alternative approach through logic and scientific analysis often seem to them to be artificial creations. The philosopher, says Wittgenstein, shows the fly how to get out of the fly-bottle, but they suspect that the philosopher has put the fly into the bottle to begin with.

Between the two world wars political philosophy provided a useful way into philosophy for the non-logician. There were real differences between the political parties; there were international conflicts which appeared to raise the deepest issues of conscience and principle; few young people were politically indifferent; rights and liberties, sovereignty and nationalism, loyalty and world revolution were common talk in café and bar. All this has gone. The main problems now seem to be those of personal and human relationships. Apart from CND, there is political apathy and a feeling of political helplessness in the face of vast and insoluble problems. So moral philosophy is now the discipline which can start from the things that matter to ordinary people.

But I have found recent books on moral philosophy (such as those of P. H. Nowell-Smith, R. M. Hare, and B. Mayo) too difficult for a complete philosophical novice, though admirable for those who have previously broken into philosophy by some other door and come to ethics late in their philosophical education.

For the beginner, Mill's *Utilitarianism*, Moore's *Ethics*, and Rashdall's *Theory of Good and Evil* (all genuine antiques) have the necessary simplicity and have previously had to serve my pupils as introductions.

This aim explains some features of the present work which would otherwise seem odd: the start from the hackneyed hedonist position, the traversing of the dead debates of the thirties, and the postponement, to the end, of any recognition of the recent revolution in ethics. This procedure seems justified, because the recent revolution was in fact a reaction against what had gone before, and is both less intelligible and less plausible if stated in isolation as itself a starting point.

References have been sparingly given in the text, but brief bibliographies are attached to each chapter.

I

INTRODUCTION

It often happens in philosophy, as in other subjects, that it is difficult to say in advance what an enquiry is about. Here only a few preliminary indications will be given before plunging into the subject itself.

We shall be concerned, then, with the meanings of certain *words*: 'good', 'bad', 'ought', 'right', 'duty', 'praise', and 'blame'. It might be said that we shall deal with these characteristics of human conduct, but this would itself be to accept a certain theory of morals. Hence the definition with reference to words is, at this stage, safer. It might be thought that, so far, we have said what is obvious. 'Good', 'bad', 'right,' etc., are obviously moral terms. But are they? They can all be used in non-moral senses. This car *ought* to go faster. Is it the *right* petrol? Or would a *good* brand be X? Should we *blame* the carburetter? And even in connection with human actions there are non-moral uses. He is a good bat, plays the right stroke, covers up when he ought; we praise his stance, blame only his caution. How are we to eliminate these non-moral uses? Again only a provisional answer is possible to a question which is still much debated.

We would start by eliminating those uses which are common to human and non-human contexts, leaving only the uses which are specifically human. For example, 'Tulloh ought to be coming round the last bend soon' is exactly parallel to 'the 4.15 ought to be coming round the last bend soon'. But there is no railway parallel to 'Jones ought to be kind to his children'. But, it might be said, all this is unnecessary. 'Moral' is a quite ordinary word which anyone can use and understand. We know that 'good' cannot be used in a moral sense of petrol or trains or cricketers. So Jones may be a good cricketer and a bad man. But it is not so easy. Good parent and bad man? Good teacher and bad man? So problems remain. But here, as often, the best initial answer to a question 'what is X?' is to demonstrate X, whether X is shrugging or skating or scalloping. I propose, therefore, to work through one or two traditional moral problems and the traditional theories about them. And I start with the most hackneyed and well-worn theory of all.

2

PLEASURE UTILITARIANISM

The key words in morals are 'good', 'right', and 'ought'. What does the pleasure Utilitarian say about them? On good, he distinguishes between what is good as a means to an end and what is good as an end. Petrol, a knife, a short cut are good if they are effective for their purposes. But is there anything that is good for its own sake? There must be, otherwise there would be no reason to choose any means. A knife is used to cut a twig, a twig to make a broom, a broom to clear leaves, cleared leaves to encourage plants, plants for their flowers, flowers for their scent—and do we stop there? For the Utilitarian (or specifically the *hedonistic* Utilitarian) only one thing is good in itself, good for its own sake—pleasure. And only one thing is bad in itself—pain. Everything else called 'good' is good because it increases pleasure or diminishes pain.

I meet a friend walking and ask him what he is up to. I may get three quite different kinds of answer. First, he may say 'I am going to the bank to get some money' or, second, 'I am taking exercise for my health' or, third, 'I like walking'. It is to be noted that in the first answer the means-and-end analysis is appropriate. Walking is a means to reaching the bank and reaching the bank a means to getting money. The stages succeed each other. But walking is not a *means* to exercise; walking is a kind of exercise, as a box-body is a kind of motor-car. And in the third case walking seems not to be a means at all. The three answers seem to suggest three different ends: having money, having health, and walking itself. But the hedonist pushes the analysis further (and obviously in the first answer it must be pushed further). 'Why do you want the money?' 'To buy a TV.' 'Why?' 'Because I get pleasure from looking at TV.' And again, 'Why do you want to keep fit?' 'To avoid disease.' 'Why?' 'Because disease is painful.' And finally: 'You say you like walking. You mean walking gives you pleasure.' And so in all other cases. This is obviously a very plausible theory, and one often maintained by people who have never heard of moral philosophy.

What then is the Utilitarian account of 'right' and 'ought' as terms to be applied to human actions? On any particular occasion the right act—the act a person ought to do—is the act which will produce the greatest possible amount of pleasure and the greatest diminution of pain. This too is a very plausible view, and it also is frequently maintained or implied by ordinary people innocent of philosophical sophistication. To spread happiness, to diminish suffering—what more need we ask of anyone? It is the fact that these theories are so popular and so natural which justifies their selection as the first to be examined here.

First criticism

This line of attack is aimed at the account of 'good'. The suggestion was that, whatever a man does or pursues, his one aim is pleasure and removal of pain. But many men would deny this. They would claim that ends other than pleasure are desired for their own sake. The miser pursues money, the thirsty man drink; the child cries for the moon; the parent

Pleasure Utilitarianism

pursues his child's welfare, the scientist truth, the martyr the triumph of his Church. Of course, it is *possible* to say on each case that it is his own pleasure that is the goal. 'He must get a kick out of it.' 'He wouldn't do it unless he thought it would please him.' But is it true? His whole heart is often set on his special goal: he does not calculate how much pleasure it will give him.

Bishop Butler was the first British philosopher to emphasise this. 'All particular appetites and passions are towards *external things themselves*, distinct from the pleasure arising from them; ... there could not be this pleasure were it not for the prior suitableness between the object and the passion.'[1] Pleasure comes when I get what I want. The pleasure could not occur unless I wanted the object. A man who gets what he does not want gets no pleasure. Butler would maintain that pleasure is *always* a by-product of the achievement of some goal other than pleasure. He goes too far here, because it is possible to do something purely for the pleasure it gives. Eating sweets, choosing a menu, are examples of this. Nevertheless Butler is surely right in general. Most men most of the time pursue different specific objects without a thought of the pleasure which (no doubt) will come if they attain their objects.

The hedonistic analysis is plausible because of the regularity with which pleasure accompanies the achievement of any object. And similarly pain accompanies unsatisfied desire. But, though it is easy to confuse a welcome result with an end, it is not necessary to do so.

Second criticism

The hedonistic analysis of 'good' turned out to be an explanation of *all* human action. But how then can it explain specifically moral or right actions? Bentham's great work begins: 'Nature has placed mankind under the governance of two

---

1. *Sermon XI* (*British Moralists*, ed. Selby-Bigge, vol. I, para. 229)

sovereign masters, *pain* and *pleasure*. It is for them alone to point out what we ought to do as well as to determine what we shall do. On the one hand the standard of right and wrong on the other the chain of causes and effects, are fastened to their throne.'[1] But it would seem to follow that every man always does what he ought to do. There is one solution of this difficulty and it is the one to which much of the rest of Bentham's work points. The act a man does is always the act in which *he thinks* at the time will bring him the greatest pleasure. The right act, the act he ought to do, is the one which *really will* bring him the greatest pleasure. It is, of course, true that 'ought' and 'right' can be so used. 'You ought not to eat lobster. You forget it doesn't agree with you.' But, on this analysis, virtue is enlightened selfishness and sin merely miscalculation of one's own private interest. This does not begin to look like a moral theory; nor this usage of 'right' and 'ought' a moral usage. And one fatal result would follow from it. On this view no one could do something wrong, knowing or believing it to be wrong. For this would mean knowingly to reject one's greatest pleasure which (on the theory we are examining) is psychologically impossible.

Third criticism

This turns on the question 'whose pleasure?' What is it that makes an act right? That it leads to the greatest amount of pleasure *for me* or *for all those affected by it*? Bentham and Mill constantly confuse this issue and repeatedly give both answers. In Bentham's definition of utility the weakness is obvious. 'The principle of utility judges any action to be right by the tendency it appears to have to augment or diminish the happiness of the party whose interest is in question . . . if that party be the community the happiness of the community; if a particular

1. *An Introduction to the Principles of Morals and Legislation* (opening sentences)

individual, then the happiness of that individual.'[1] But this will not do. Bentham must choose. Only the general happiness would seem a possible basis for a moral theory; and this might have served as an alternative solution to the paradox in Bentham that pleasure and pain determine not only what we ought to do but what we shall do. My own pleasure and pain determine what I shall do; but the greatest happiness of the greatest number determines what I ought to do. Both of these propositions (as we have seen) are plausible. But unfortunately they are incompatible with each other. Utilitarianism owes its perennial attractiveness to the belief that it is possible to ride these divergent steeds side by side. But if I necessarily do pursue only my own pleasure it is impossible for me to pursue the general happiness and so impossible for me to do what I ought to do.

The solution proposed both by Mill and by Bentham for this difficulty is the use of sanctions. I am necessarily selfish so why should I bother about the feelings of others? Bentham replies: 'If you do not, the law will catch up with you and you will not like that' (political sanction), or 'If you do not you will be shunned and boycotted and you will not like that' (popular sanction) or 'God will punish you hereafter' (religious sanction). Mill adds: 'You will not beat your children because you have parental feelings and you will be unhappy if they suffer; or you have sympathetic tendencies and other people's happiness will make you happy' (internal sanctions). But there are two difficulties about such solutions. First the sanctions are not certain. 'Crime doesn't pay' only for the criminal who is caught. The NSPCC is kept well occupied by parents who can be quite happy when their children suffer and even because they suffer. And secondly the sanctions again reduce morality to enlightened selfishness, which is not moral at all.

1. *British Moralists* I, paras. 359, 360

Fourth criticism

This criticism applies whether one adopts private pleasure or the general happiness as one's criterion. The great claim of Bentham and Mill (as of all other Utilitarians) is that their theory provides a single objective standard of right and wrong. Bentham uses the word 'calculate' and the criticism now relevant is aimed against the notion of a calculus of pleasure. Most people would agree that a pleasure (or pain) may vary in quantity. A drink gives more pleasure when one is thirsty and some kinds of drink give more pleasure than others. But, for calculation, this is not enough. How much more pleasure? Twice as much? Or thirty per cent more? This is bad enough; but there are greater difficulties still. Bentham says, rightly, that there are various factors to take into account in estimating pleasures; intensity, duration, fecundity (i.e. productivity of further pleasures), purity (i.e. freedom from pain). But now how are we to measure one of these against another? How long has a mild pleasure to last if it is to balance an intense but brief pleasure? It is obvious that these questions are unanswerable. And there is a final difficulty of the same kind. 'Mankind is under the dominion of *two* sovereign masters—pleasure and pain.' But how are *they* to be compared? There would have to be some third entity, common to both and applicable as a standard. It seems clear that the Utilitarians thought of the pleasure-pain scale as if it was a single scale with pleasure as plus and pain as minus, like a temperature scale. But this is clearly a mistake. On a temperature scale—if it is to allow measurement—what we call hot and cold must both be degrees of heat and the position of the scale zero is arbitrary (it differs in Centigrade and Fahrenheit) while the true zero is the unattainable absence of all heat at the lower end of the scale. But the zero in pleasure-pain is freedom from both, as in unconsciousness. So here too calculation is impossible.

Fifth criticism

A final problem for the Utilitarian is our use of 'good' and 'bad' in describing motives and intentions. We might agree with him that a man did the *right* thing in leaving money in his will to help on medical research, resulting in the decrease of human suffering. But if he did this from the motive of spite and with the intention that his children should not get his money, we should say this motive or intention was bad. The logical line for a Utilitarian to take is that taken by Bentham. We call motives bad when in most cases they lead to bad results, though not in this; and we call the intention bad because the result aimed at is bad. But these answers seem unconvincing.

REFERENCES

J. Bentham in *British Moralists*, ed. Selby-Bigge
J. S. Mill, *On Utilitarianism*

Criticism of the theory that pleasure alone is desired:
H. Sidgwick, *Methods of Ethics*, i, pp. 31–40

Criticism of measurement of pleasure:
Sidgwick, ibid., i, pp. 109–35

General criticism of pleasure utilitarianism:
G. E. Moore, *Principia Ethica*, ch. iii

3

IDEAL UTILITARIANISM

Among the difficulties of pleasure Utilitarianism were the following:
   (a) men value ends other than pleasure;
   (b) are we to pursue our own good or the general good?
   (c) what value can be attached to motives and intentions?
All three are met by the type of Utilitarianism we shall now consider, the 'Ideal Utilitarianism' of G. E. Moore.

(a) *Ends other than pleasure*

Moore holds that aesthetic enjoyment and personal affection are good in themselves. This is clear to him, first because either of them by itself without any other accompanying or resultant good would be considered by all as worth having or bringing about, and second because the addition of either to any state of affairs would make the total worthwhileness of that state of affairs greater. And, in connection with the latter point, Moore notes that the word 'addition' may be misleading. Pleasure, for example, as such by itself, has little value, but pleasure associated with aesthetic experience greatly enhances the total value of the state of affairs. Knowledge by itself has little value but the knowledge associated with personal affection ('human understanding') greatly enhances the total value. This is Moore's principle of 'organic unities' by which the value of a whole depends on the relationships of the parts and is different from the values these parts would have if they were separately estimated and simply added.

(b) *Private or general good*

Moore is clear that the good results which make actions right must include the results for people in general and not for the agent only. With characteristic accuracy he insists that all that counts is the total quantity of good. The Utilitarian principle 'greatest good of the greatest number' may be self-contradictory. For it may be possible to produce a greater good in total by restricting it to a smaller number of people. Or again Utilitarians speak of the general happiness or the good of the community. But who are included in 'general' or in 'the community'? Again it is obvious that the only logical answer is to reject any limitations (to a particular nation or society), but yet not to say 'all mankind' when some of mankind will be wholly unaffected. The accurate formula would be 'the greatest total good for all those affected by the action in question'.

(c) *The value of motives and intentions*

Moore maintains that an act can be judged *right* or *wrong* without any reference to motives or intentions. But, besides passing judgment on the act as right or wrong, one can reach quite a different judgment on the agent as morally praiseworthy or blameworthy. To this second judgment motives and intentions are relevant. Moore is right in drawing attention to this dual nature of moral judgment, a duality frequently emphasised but also frequently overlooked. 'The surgeon did the wrong thing but you couldn't blame him as he couldn't have known the patient had this allergy.' For confusions one need only look at judgments on great political decisions such as the Munich Agreement. It is clearly possible to hold that it was right to make the agreement, because if we had fought in 1938 we should have fought without the Commonwealth, and without the Hurricanes and Spitfires which won the 'Battle of Britain'; but that Chamberlain deserves no praise for it, because these were not his intentions and motives. Or, vice versa, it may be said that it was the wrong decision because if we had fought in 1938 we should have had Russia and a strong Czech army and frontier to help us, but that Chamberlain was to be praised for the decision because his motive was an overwhelming desire for peace.

Now clearly these developments by G. E. Moore are an immense advance on pleasure Utilitarianism both in clarifying it and in turning it into a defensible moral theory.

Criticism of Moore's theory

Moore himself noticed that the main alternative moral theory to his own was the theory that acts are right when they exemplify certain moral rules (such as the Ten Commandments). Against this view he urged that every moral rule has exceptions and these exceptions occur precisely when keeping the rule would do harm or breaking it would do good. Such moral rules, he thought, are adopted and followed because their implementation usually produces good results and thus a man can save time by not having to think out the results of every action.

Moore did not notice that there are other kinds of exception to moral rules, which arise when two rules conflict. I am told something in confidence and then asked a question which I can answer truthfully only by breaking the confidence. But these exceptions would reinforce the case against binding and absolute rules. In fact since *any* two moral rules may conflict it is possible to hold a theory of absolute rules only if you have only *one* absolute rule. Some pacifists come near this position. Their absolute rule prohibits taking human life (or

Ideal Utilitarianism

using violence) and any other rule or moral obligation which conflicts with this is overridden by it. As against a theory of absolute rules, then, Moore's arguments are effective.

But there is a version of the 'Rules' theory which avoids these difficulties. It is that defended by H. A. Prichard and Sir David Ross. A theory of absolute rules has to maintain that a rule must be kept no matter what the consequences. '*Fiat justitia, ruat caelum.*' Prichard and Ross agreed that if the consequences of keeping a rule were *sufficiently* bad or of breaking it *sufficiently* good it should be broken. And they held that a rule was binding only if no other rule was involved and if two rules conflicted the more stringent rule should prevail. They observed that the reference to consequences was equivalent to adding rules requiring the production of good and the diminution of bad results. They were thus not anti-Utilitarian —their moral system included a Utilitarian element. They were left with two problems. (a) How are rules themselves to be justified? (b) How is a decision to be made when rules conflict? They answered the first question by saying that justification was unnecessary. It is just self-evident that promises ought to be kept and the truth to be told, as mathematical truths are self-evident. This and other answers to this first question are considered in this next chapter. In answer to the second question Ross says that cases of conflict are to be settled by each individual's judgment of the relative stringency of the claims upon him. No principle for such decisions is available. For if there were such a principle it would be the one absolute duty, which is just what the Utilitarians thought their principle was.

This case against Utilitarianism then has to depend on showing that keeping a promise and telling a truth are right not because of their consequences, not because of the good they will do. Here the critics appeal directly to the ordinary man's moral experience. If I owe my tailor ten pounds and I have ten pounds it will not do to say 'How can I do most good with this money?' or to consider that it will do more good if I give it to Famine Relief than to my creditor. If anyone does not *see* this there is no way of proving it to him. I can try to clarify the situation by saying: 'But it's really your tailor's money. Charity is not a virtue when you give away other people's money.' But Robin Hood and socialist governments think it is. It is obvious that, so far, Ross and Prichard do express ordinary moral standards better than the Utilitarian. The normal Utilitarian explanation of examples such as the debt example is to appeal to the indirect consequences of breaking promises or not paying debts. These actions do not merely deprive people of goods; they deprive them of *expected* goods; they cause disappointment as well as loss.

To meet this, examples may be cited in which no disappointment arises. I promise A to do something for B who does not know about my promise, and is therefore not disappointed if I break it. But A knows, and my failure will shake his confidence in promises. Not to pay a debt deprives a creditor but also weakens the credit system. To meet this sort of case, Ross devises an example in which *no one* will know the promise has been broken. 'If we suppose two men dying together alone, do we think that the duty of one to fulfil before he dies a promise he had made to the other would be extinguished by the fact that neither act would have any effect on the general confidence?'[1]

The reaction of many readers to such an example is to say that 'desert-island morality' cannot be quoted in evidence. Professor Nowell-Smith says, for example: 'I confess to being quite unable to decide *now* what I should say if a desert-island situation arose. Moral language is used against a background in which it is almost always true that a breach of trust will, either directly or in the more roundabout ways

1. *The Right and the Good*, p. 39

which Utilitarians suggest, do more harm than good; and if this background is expressly removed my ordinary moral language breaks down.'[1]

I admit that when I first read Ross my reaction was similar to this and other critics must have urged a similar misgiving. For Ross, in his later book *Foundations of Ethics*, produces obviously real-life examples to illustrate the same point. I too kept an eye open on my own experience and within a few years I had come across half a dozen such cases. The essential point in all these cases is that no one should know the rule has been broken and so there can be no effect on the general confidence, etc. I have quoted some of these cases elsewhere,[2] so one example will suffice here. A former pupil of mine told me he had become secretary to a very rich man. He asked his employer what should be done with begging letters and was told: 'Put them in the waste-paper basket. We have no time to verify them all, and you know the list of my charities, which I have thought out with care.' The employer had a habit of stuffing a roll of bank-notes into the pocket of any suit he was wearing and his secretary was constantly extracting these bundles from suits being sent for cleaning. He handed them to his employer who always put them in his pocket uncounted. One morning, having nothing to do, my friend looked through the begging letters before destroying them. One was a winner—fully authenticated and making a good case for an immediate need of £10. My friend had just fished a bundle of fifty-seven pound notes out of a pair of tennis trousers. I said, 'Well, did you send the £10?' 'No!' 'Why not?' 'It wasn't my money.' Now this is clearly not a Utilitarian reason. And the secretary had every reason to believe that no one would know that he had broken his

---

1. *Ethics*, p. 241
2. Cf. 'Punishment', *Mind* XLVIII (1939), p. 156; 'Moral Rules', *Proceedings of the British Academy* XXXIX (1953), p. 103

employer's instructions. And so general trust and confidence would have been entirely unimpaired.

There is a last ditch for the Utilitarian, and in his second book Ross foresees it.[1] One person does know that the undertaking has been broken and that is the man who has broken it. What about the effect on him, on his character? There are two effects which may be alleged. It may make him a morally worse man; and, if virtue is an end of intrinsic value, then the act is wrong as damaging that value. Or it may make him more likely to break other promises in future and while *this* broken promise has no direct bad consequences, they will.

Both these arguments, however, fail. A man's character can be damaged only by doing what he believes to be wrong. A Utilitarian must therefore have other reasons than this for believing it wrong to break a trust. The argument is circular. The other line is no better. To break a promise in order to do good will indeed weaken a man's habit of keeping promises. But no one, except a rigid Kantian, thinks that all promises are to be kept no matter what—and if he does he is no Utilitarian. As a parallel let us take the rule, sometimes given to beginners at whist or bridge: 'Third player plays high.' Thus, suppose I hold the King of spades and the first two players have played low spade cards I should play my King. (One reason being that if the whereabouts of the Ace are unknown the odds are two to one against the fourth player holding it.) But now suppose the first player plays the two of spades and the second player the Ace what do I do? What would we think of my bridge tutor if he said: 'You should play your King because the general rule is a good rule—third player plays high—and if you do not play your King you will weaken your habit of playing high at third player, and *in other cases* you will be liable to lose tricks.' Surely the answer is that the rule is a good rough guide but one to be scrapped when its appli-

1. *Foundations of Ethics*, p. 104.

cation would be obviously useless. It would be a bad thing to get into a *rigid* habit of playing high as third player. If this is to be a habit it must be a flexible habit capable of being broken on suitable occasions. So too for a Utilitarian moral rules are rough guides to good results—to be broken when better results will accrue; and the case in question is precisely one of those.

So I conclude, with Ross, that the Utilitarian cannot explain our conviction that it is frequently right to pay a debt or keep a promise when we could do more good by using the money for some other purpose or by breaking the promise.

Ross sums up his attack on Utilitarianism by saying that it 'fails to do justice to the highly personal character of duty. If the only duty is to produce the maximum of good, the question who is to have the good—whether it is myself, or my benefactor, or a person to whom I have made a promise to confer that good on him, or a mere fellow man to whom I stand in no such special relation—should make no difference to my having a duty to produce that good. But we are all in fact sure that it makes a vast difference.'[1]

This reference to the 'personal character' of duty and to 'special relationships' seems to me, however, to conceal the fact that these exceptions to the Utilitarian position are of two very different types and involve rejecting two quite different corollaries of the Utilitarian position. And, as ordinary people sometimes confuse these two, it is worth distinguishing them. It is sometimes my duty to do something for someone because of what is rightly called a 'special personal relationship'—for example, my parents, my colleagues, my pupils. This offends against the inevitable corollary of the Utilitarian principle that, provided the good is produced, it does not matter who has it —or, as Bentham put it, everybody is to count for one and nobody for more than one. But it is also sometimes my duty

1. *The Right and the Good,* p. 22

to do something because of some past occurrence; promise-keeping and debt-paying are the obvious cases here. The corollary of Utilitarianism against which these cases are fatal is that the past can never affect the rightness of a present action. Only the future counts because acts are made right by their results. Of course such duties involve other people (creditor and promisee) but their special point seems to me to be lost if they are described as depending on special personal relationships. I have no 'special personal relationship' to Messrs Dodson and Fogg to whom I owe money. Indeed I do not know who they are. They may have been 'taken over' years ago.

To see how ordinary people are sometimes misled here, two examples will suffice. One sometimes hears that a man's duty to his parents rests on what they have done for him—here a personal bond is confused with a debt. But it is clear that this is not the whole truth. No doubt when my parents have done a lot for me, I owe them a double duty as parents and as benefactors. But if gratitude were the whole story then what would happen to the duty of parents to children? It would rest on a lively expectation of favours to come. Similarly if it is asked why I should cherish my wife, I am sometimes told that this is because of my marriage vows. Here again a duty to a person depending on a special relationship is confused with a duty (that of promise-keeping) depending on the past. It is clear that this answer too is no good because, if the vows were the reason, then husbands married by ceremonies which include no vows would have no duties to their wives. No doubt I (having made my vows) have a double duty, both to my wife and to my word.

REFERENCES

On Ideal Utilitarianism:
    G. E. Moore, *Principia Ethica* especially chs. v, vi; *Ethics* especially ch. v
    W. D. Ross, *The Right and the Good*, chs. i, ii

H. A. Prichard, 'Does Moral Philosophy Rest on a Mistake?' in *Mind* (1912), reprinted in *Moral Obligation*

On 'desert island' examples:
J. Narveson in *Analysis* (January 1963)

4

MORAL RULES

(1) *Self-evidence*

We have seen how the criticism of Ross and Prichard against Ideal Utilitarianism is based on the acceptance of certain moral rules and how their theory avoids the difficulties of a Kantian commitment to absolute rules. It is natural to ask how such moral rules can be justified. The view maintained by Ross and Prichard was that such rules are self-evidently valid. In some cases this self-evidence is achieved by the words used. 'Debts ought to be paid' raises the question of the meaning of the word 'debt'. If it means 'moneys owed' then, of course, the money ought to be paid. In the authorised version of the English Bible (1611) the ten commandments in Exodus xx include 'Thou shalt not kill'. Now here there is nothing in the word 'kill' to make this tautologically self-evident, like 'moneys owed ought to be paid'. And, of course, the difficulty arose that the Old Testament is full of killings of which God is held to show no disapproval. So in the Revised Version (1885) the commandment reads: 'Thou shall do no murder.' But now what does 'murder' mean? Does it not mean those killings which are reprehensible, which ought not to be committed? It might be replied that murder can be defined as killings which are not intentional, not carried out under legal warrant, not in pursuance of a military duty, etc. But it is surely true that, alongside all this, 'murder' has a morally condemnatory sense and therefore the Revised Version commandment is a tautology.

It is clear, however, that some rules are not tautologies—'Tell the truth', 'Honour thy father and mother'—and the problem of validating these still remains. The difficulty for Ross is that not everyone recognises the moral rules he does. Ross meets this by saying that such people have not reached a sufficient degree of maturity. 'Three threes are nine' is self-evidently true, even though its truth is not evident to a savage who cannot count beyond five. While this answer will explain why some men do not see the truth of moral propositions which are self-evident to Ross, it does not explain why other men have rules of which Ross would disapprove. The variation in moral rules from people to people and man to man seems to rule out a self-evidence theory.

In his later work Ross meets this by distinguishing between basic rules and dependent or derivative rules, which result from applying a basic rule to special circumstances. Ross holds that apparent differences rest on agreement on basic rules. The difference arises either from the special circumstances in which the basic rule is applied or from different factual beliefs entertained by the people in question. Thus the duty of a Briton to help the police to arrest a murderer and the duty of a Sicilian to kill a member of the murderer's family are basically the same duty of requiting murder or deterring murderers applied to different sets of circumstances, those with and without an effective central legal authority. The early fathers of the Church explained the polygamy of the Jewish patriarchs by the fact that the Jews had a small population and were surrounded by hostile peoples who constantly took toll of their adult males in war. It was essential that the birth-rate should be maintained.

There are equally obvious examples of cases where duties which seem to us very odd are explicable by the *factual* beliefs of the people concerned. The tribesman from northern Siberia when he kills a woman who steps across his shadow shares with us the basic (and self-evident) moral belief that deleterious influences should not be allowed to damage the souls of men, along with the mistaken beliefs that a woman is an inferior creature and that his shadow is part of his personality. Other tribesmen who kill and eat their parents show both the features noted above. They agree with us that we should be kind to our parents. But, as they are nomadic tribes who have to rely on fleetness of foot to escape enemies, animal or human, the kindest thing is to kill them. As for eating them, they agree with us that we should honour our parents and try to acquire their virtues. They combine this with the dietary belief that men come to resemble what they eat.

But it is not possible to reduce all differences in moral rules to variations on agreed themes. The conscientious objector and his opponents are not agreed on fundamentals nor are those who think that suicide and divorce are sometimes right and those who hold they are always wrong.

(2) *Empirical generalisations*

Those who for these reasons or others reject the self-evidence of moral rules are liable to go to the opposite extreme and hold that moral rules are empirical generalisations resting on particular moral judgments, which are ultimate. A number of particular actions are recognised, each independently from the others, as being right. It is then noticed that all of them have some other similarity in addition to their rightness and it is then inferred with a certain degree of probability that rightness and this other characteristic will accompany each other always or in the next observed case. This was the view of

Adam Smith. 'General maxims of morality are formed, like all other general maxims, from experience and induction. We observe in a great variety of particular cases what pleases or displeases our moral faculties, what these approve or disapprove of, and by induction from this experience we establish the general rules.'[1] This theory is echoed, though without Adam Smith's clarity, by other moralists right down to the present day. C. L. Stevenson, for example, holds that the attachment of a moral adjective each as 'right' to a *kind* of action is merely the result of 'habit and rough generalisation'. It is due to 'the psychological economy that comes from ordering of attitudes in some sort of classification'.[2]

Thus in order to establish that acts having a certain characteristic X are right it is necessary first to recognise that they are right independently of their possessing this character X, and then to observe that they also have the character X, and only then to conclude that rightness and the character X are connected. These recognitions must be independent. For, if the agent approved of giving a book to Jones *because* he had borrowed it or disapproved of sticking a pin into Smith *because* it would cause pain, the connection between right and the return of borrowed articles or between wrong and causing pain would not have been an inductive generalisation empirically established. It is this feature, concealed in most statements of it, which inclines me to hold that it is indubitably false.

The existentialists would seem to come near this view with their insistence that actual choices determine our valuations and not vice versa. But their examples show that the valuations come first. In an example given by Sartre a young man is torn between the claims of his mother who, deserted by her husband and bereft of her other son, lives only in him, and the call of

---

1. *The Theory of Moral Sentiments*, 1st ed., p. 502 (*British Moralists*, ed. Selby-Bigge, vol. i, para. 344)
2. *Ethics and Language*, p. 95

the Free French movement in England which would mean leaving his mother. But the anguish of the choice is due to the fact that these claims precede it. It is true as Sartre rightly stresses that there is no third principle available to guide the choice between these two. Therefore, in a sense, it is not clear *what* value he places on devotion to his mother until he has made the choice, not clear even to him. But this throws no doubt on the concept of moral rules, as Ross interprets them, as claims on the individual. Another way in which it can be seen that the basic moral judgments cannot be concerned with particular acts is to ask what the description of a particular act would be. Helping at that car smash in the High Street last Tuesday, or paying that £5 to Brown. But the full description would include the number of the car, the clothes I was wearing: the writing of a cheque and enclosing it in an envelope and . . . and . . . Of course you will say all this is irrelevant. But how is it to be decided what is irrelevant? One remembers the cricketer who

> Bowled twenty-one wides in an over,
> Which had never been done
> By a clergyman's son
> On a Tuesday, in August, at Dover.

Relevant features of a particular situation are precisely those which, being general, fall under some rule or other.

It may be suggested that Utilitarianism holds that moral rules are empirical generalisations. We discover that several actions are alike in that each produces the maximum good possible in the circumstances. We find they also resemble each other in being the keeping of promises or the telling of the truth. We conclude that promise-keeping or truth-telling will be likely in most cases or the next case to produce good and therefore to be right. But this is not a pure case of the theory,

Moral rules

for the empirical generalisation has to be supplemented by two other general statements before it is relevant to moral conduct at all. First the statement that certain *kinds of result* are good and second that an action is right if it produces more good than any possible alternative action. Neither of these is an empirical generalisation. It follows too that the empirical generalisation is itself not a moral generalisation but an ordinarily factual generalisation asserting that certain kinds of actions have been found generally to conduce to pleasure or whatever other result the first supplementary principle holds to be good.

(3) *Utilitarian precepts*

We may now consider on its merits the Utilitarian view analysed above. We have seen that, on this view, it would follow that where an action exemplifies a moral rule but would have less good consequences than an action breaking the rule a Utilitarian would seem compelled to say the rule should be broken. We have also seen how Utilitarian attempts to evade this conclusion break down. G. E. Moore when faced with this type of case reaches a very odd result. He admits that we believe it right to keep a rule in cases where we believe more good would be done by breaking it. He holds the first belief is justified but only because the second is likely to be mistaken.

For if it is certain that in a large majority of cases the observance of a certain rule is useful, it follows that there is a large probability that it would be wrong to break the rule in any particular case; and the uncertainty of our knowledge both of effects and of their value, in particular cases, is so great, that it seems doubtful whether the individual's judgment that the effects will probably be good can ever be set against the general probability that that kind of action is wrong. Added to this general ignorance is the fact that, if the question arises at all, our judgment will generally be biased by the fact that we strongly desire one of the results which we hope to attain by breaking the rule. It seems, then, with regard to any rule which is *generally* useful, we may assert that it ought *always* to be observed, not on the ground that in *every* particular case it will be useful, but on the ground that in *any* particular case the probability of its being so is greater than that of our being likely to decide rightly that we have before us an instance of its disutility.[1]

Now this is a very queer theory indeed for a Utilitarian to have to adopt. Moral agents are now subject to rules of Kantian rigidity, rules with no exceptions. In the hackneyed example I must tell the murderer which way his victim went because truth-telling generally has good consequences. I must never break a date to help at an accident even if I am a doctor because promise-keeping generally has good consequences. Moreover the supporting arguments used by Moore are odd. The crucial cases for morals arise over 'conflicts of duty', and, in these, desire is not normally strongly engaged on one side. I have no violent desire to help Dr Barnardo's Homes rather than my tailor.

We return to the difficulty mentioned earlier that different people have different rules. Other men have felt about *harakiri*, about duelling, and about vendettas just as we feel about debts and promises. We feel it wrong to ask what good it will do to pay the debt. So they would feel it wrong to ask what good will result from my committing ceremonial suicide or calling out my traducer or killing a brother of my brother's murderer. There is no question of doing good; it is a plain duty; honour not utility is at stake.

Faced by this difficulty some may sit back contented with a subjectivist or relativist view, as they would with differences of judgment of beauty. They would say that it is just a matter of individual taste and needs no explanation, any more than differences over oysters or caviare.

1. *Principia Ethica*, p. 162 (italics original)

Others will urge a causal explanation. We approve of debt-paying and the Sicilians of vendettas because we and they have been brought up so to approve. Rules are accepted because society has indoctrinated men. Get a child young enough and you can give it moral scruples about anything whatever. This, however, cannot be the whole truth. If it were, it would be impossible for a man ever to question the rules of his society and therefore impossible for its rules ever to change (unless the man came like the Prince Consort, bringing into a society the rules of some other society in which he had been brought up, or one society was forcibly absorbed into another as Latvia and Estonia were during the last war, with consequent changes in their moral codes).

It is clear, however, that moral rules can be criticised and sometimes changed by the members of a society whose rules they are. So they should be capable also of being defended. How can this be done? I think by using Kant's test, but by giving it a Utilitarian twist which he tried to avoid. To find whether a rule is justified, Kant asks 'what if it were universalised?' In some cases universalisation—what would it be like if everybody did this?—produces a self-contradiction and then we know the act is wrong for no rule is possible. A lie is a false statement made when it will deceive a hearer. A fairy story is not a lie because it will not deceive the hearer, for he is not expecting the truth. But if everyone said what was false, no one would expect the truth and so no one could be deceived. Hence 'universal lying' is a self-contradiction. Stealing is *appropriating* the private *property* of another. But, if everyone took what was in other people's keeping, there would be no private property and 'appropriation' would be impossible. So 'universal stealing' is a self-contradiction.

In other cases, however, Kant has to admit that he cannot will the action to be universal not because of self-contradiction but because we would not welcome the results. There would be no self-contradiction if all men agreed to let their talents atrophy; but, says Kant, I cannot will this universally because I cannot will that my own talent should be allowed to lie unused.

Kant has been accused of Utilitarianism in using these arguments as if he had said that the world would be a worse place if all men lied or stole or refused to cultivate their talents. This is not so. But the self-contradiction argument leaves a gap. Why should the institution of communicative language or private property be accepted at all? Why should we not use language (as the Irish are sometimes said to do) to please or amuse rather than to inform? There is no logical impossibility about a lotus-eating society in which no one develops his talents. How would it be, then, to add the Utilitarian criterion? A rule is acceptable if a society in which it was observed would live a better life than one in which no rule, in this field, or some other rule were accepted. This solution seems first to have been suggested by Francis Hutcheson. 'The way of deciding about any disputed *practice* is to enquire whether this conduct or the contrary will most effectively promote the public good. The morality is immediately adjusted when the natural tendency or influence of the action upon the universal natural good of mankind is agreed upon.' And compare his remark on passive obedience: 'The point disputed among men of sense was whether *universal* submission would probably be attended with greater evils than temporary insurrections where privilege were invaded.'[1]

I wish to emphasise the difference between accepting the Utilitarian validation of a rule and accepting the Utilitarian justification of a particular action. As this is crucial and often missed or obscured, it has to be made clear. Even G. E.

---

1. *An Inquiry concerning the Original of our Ideas of Virtue or Moral Good*, sect. ii, para. iii, *British Moralists* (ed. Selby-Bigge, vol. i, para. 112) (my italics)

Moore—usually so meticulous—obscures it when he says, 'Apart from the immediate evils which murder generally produces, the fact that if it were a common practice the feeling of insecurity thus caused would absorb much time which might be spent to better purpose is perhaps conclusive against it.' What is not clear here is whether Moore is judging a particular murder to be wrong, on the grounds that, *if* murder were general, the results would be bad. This is indeed the theory under review in the present section. I judge a particular act to be right though its consequences may be less good than those of some alternative act open to me. It is right because it exemplifies a certain rule. But I judge the rule to be right because *if* it were generally observed the consequences would be good. This is often how what we call 'acting on principle' is in fact defended. Two hundred years ago I might have refused to challenge to a duel a man who had pushed me off the pavement. Yet I am the better swordsman and he a man of whom the world would be well rid. Asked for my reasons for refusing I should not appeal to the consequences of the particular duel (for they would be good). Nor should I cite the consequences of my refusal on the general system of duelling. This is the 'indirect' argument examined earlier (p. 26). For I should still claim I was right to refuse even if I was regarded as a crank or a coward and the system remained unshaken or even fortified by my refusal. I should say that duelling is a bad system. Its adoption makes all injuries equivalent and all a matter of life and death. This clearly could not be a reason against fighting any particular duel.

It is interesting, as an example of the variations in persuasive language, to see how this same point about the system can be described in opposite ways. Hegel speaks of 'the barbarity of the formal code of honour, which found in every injury an unpardonable insult'. Burke refers to 'that sensibility of principle, that chastity of honour, which felt a stain like a wound'. Moreover the duelling system considered as a reaction against wrongs and injuries has absurd consequences. It makes punishment depend on aim or swordsmanship, not on guilt, so that a bully who is good with his weapons would be able (and would be encouraged by the system) to go about insulting people and counting on his skill to get him out of trouble. This argument too cannot be used in my particular case, when I refuse to fight though I am the better swordsman and the bully would not escape punishment.

What can be said against this validation of moral rules by their consequences? It has been objected[1] that nobody who keeps a promise or pays a debt does so for such recondite reasons. It is of course true that, in life as in games, many rules are obeyed by people who have no idea why the rule holds. And 'why' here demands not history but justification. Most of our beliefs in all fields including ordinary matters of fact are held with no reflection on their justification. But this does not mean that there can be no evidence for them. Moral life like all ordinary life is too short for everyone to be constantly considering or obtaining or demanding adequate evidence for all his beliefs. But when a belief is challenged one may ask what would be the right kind of evidence and have some idea where to look for it.

Second, it has been objected[2] that a man has a duty to pay his debts even if he regards the system whose rule this is as inferior in its consequences to other possible systems such as barter or loans only on security. Hence the rightness of his payment cannot be derived from the superior good produced by the system. There are two answers to this. First there is the duty of supporting the existing system even if it is inferior because of the good done by such support in maintaining trust and confidence. Second, the man himself accepted the system when

1. I. Gallie, 'Oxford Moralists', *Philosophy* (1932)
2. N. G. H. Robinson, *The Claim of Morality*, p. 277

he incurred the debt. People who disapprove of credit systems should not incur debts. A parallel case is that of a judge who enforces a law whose general results he deems to be bad, because of the good done by law enforcement in general, and because he accepted this set-up when he took up his appointment. He could resign.

It may be objected that I have now myself reduced moral rules to empirical generalisations—the view I rejected earlier. For if an action is justified by reference to a rule and the rule is justified by the fact that its general observance would have good results how else could this be determined? Must it not rest on the inspection of repeated applications of the rule compared with the results of observed applications of other rules or no rules in the field in question?

Here the parellel with rules of a game may help. Some of the rules of a game may be called 'constitutive'. They determine what kind of a game it is to be: that each bridge player should have thirteen cards; that the association footballer should not handle the ball. It might be said that there is no question of justifying a rule of this kind. If you don't like the rules, don't play this game. Changing the rules would simply result in a different kind of game, such as we owe to Mr W. W. Ellis who in 1823 'with a fine disregard for the rules of football as played in his time first took the ball in his arms and ran with it'. But it could well be argued that the new game is a better game, as it has been by Mr Ellis's sectaries; and as most would agree that bridge is a better game than whist and contract than auction bridge. So even these rules can be justified by their consequences. Then there are other rules which can be altered without changing the whole character of the game like the revoke rule in bridge or the offside rule in football. These may be called 'regulative' rules. But few of these rules, whether 'constitutive' or 'regulative', could plausibly be called empirical generalisations. They are not justified by observing, prior to the formation of the rule, a number of games and discovering that those in which a certain practice was followed were better than those in which it was not and then laying down the practice as a rule. (This is not to deny that such experiments are possible, and there have been some recent cases in Rugby football.) But it does not require repeated experiments to establish that the reduction of services in tennis from two to one would diminish the present advantage of the server and hence the premium placed on height and strength. No repeated observations are needed to show that a check on bodyline bowling would improve the spirit of cricket or that the change from auction to contract would reduce the element of luck in bridge. As in games, so in life. No repeated observations are required to establish that loans on security will be given more readily than credit without it, or that the duelling system encourages the bully who is a good swordsman or shot. It is indeed one reason for some current suspicion of sociology as an empirical science that it goes to much trouble to establish by empirical methods conclusions which are obvious without them, such as that the children of divorced parents tend to be emotionally disturbed, that two juries faced with the same evidence may reach different conclusions, or that a visible luxury, such as a car or a television apparatus, will become a 'status symbol' and will be bought by some people who cannot afford it and by others who seldom use it.

Moral rules vary in the degree to which they are constitutive or merely regulative. There are some rules without which no civilised society could survive and few values could be achieved. The rules against killing and promise-breaking are of this kind (and this may be the reason why they have been supposed to be self-evident). Then there are other rules essential to a particular institution or a particular kind of society such as the rule against theft. Its abolition would involve a complete change—for example to a communist society. Or the abolition of the rules making parents responsible for their children would mean a change to a Platonic Republic in which children were taken from their parents at birth and brought up in government crèches and nurseries. These other societies too would have their own constitutive rules in place of those against theft and parental neglect; the communist state a rule against sabotage and the Platonic Republic a rule that children should not know who their parents are. Then there are regulative rules which are alternative to others within a given system and alterable without destroying society altogether or changing it to a completely different set-up. The age at which children cease to be the responsibility of their parents may be altered or a property system may rest on entailed inheritance or free testamentary disposal. But in regard to all these alternative institutions, or rules within institutions, it is always possible and legitimate (and sometimes necessary) to ask which are the best rules and this means asking what consequences follow or would follow from their observation or imposition.

A further complication in considering a rule as an empirical generalisation based on observing the good consequences of a series of similiar particular actions arises when the rule is one of a number or rules essential *together* for the working of an institution so that the rule cannot be said *alone* to have good results, since any alteration in it would involve altering other rules. If we defend the opening of all professions to women, we shall find (as Plato did) that we shall have to alter our views about the duties of mothers to their children. If we defend equal pay we may find ourselves forced to defend family allowances.

In estimating the effects of having a certain rule there is a further consideration which again makes 'empirical generalisation' seem an inadequate description. This description would require that the rule be kept in a large number of cases and from each case taken by itself (or from each of a majority of all cases) good results be seen to follow. But normally the good results of a rule do not follow seriatim on the several observances of it. They follow from *awareness* that the rule has been adopted. A rule is laid down that thefts will be punished. The best results are achieved not when a series of people are punished for theft, but when no one is punished at all, when the threat succeeds. The good produced by a rule that loans will be repaid is produced by the *belief* that they will be repaid; for the good in question is the availability of resources in time of need. Actual punishments or repayments come in primarily as applications of the rule and not as sources of good (though they may incidentally achieve some good too).

This point may be seen in another way. If the good produced by a rule were produced by the individual actions of keeping it, one would expect the good to vary, at least roughly, with the number of observances and the harm done by breaking it to vary with the number of breakages. But this is not the case. A few failures to repay or to punish may produce no harm at all. People feel no less secure because a few more thieves escape justice. They find it no less easy to get credit when the few bad debts which any tradesman expects rise from one to two per cent. But as the breakages rise there are two successive danger points, one when confidence is shaken and another when the system breaks. When the first point is reached people who have previously paid no special attention to the rising tide of failures to pay or to punish suddenly take special precautions, looking up customers in the local directory before giving them credit or double-locking their doors and taking their valuables to the bank. Then, if the failures mount further, a new crisis comes when the system collapses altogether; tradesmen stop giving credit to anybody or citizens arm themselves and organise posses of vigilantes. Banks know these two danger points. Ordinarily there is no problem; deposits balance withdrawals. Then as confidence weakens and saving ceases owing to a threat of inflation withdrawals may rise. And, at a certain point, measures may have to be taken to counter this—restricted credit or a rise in the bank rate. But if withdrawals still increase there may come a new crisis of confidence, a run on the bank and a collapse of the system.

These successive crises of confidence may indeed be largely independent of the number of breakages of the rule. Confidence and the loss of it are infectious. A whole series of thefts at a railway terminus will not shake confidence because one victim rarely knows that there are others, while far fewer thefts in a single local area will make a substantial difference. A dozen murders widely distributed and unpublicised will shake confidence less than three given front-page treatment or restricted to one locality.

Another corollary of the importance of confidence is that a bad rule may produce good results simply because it is a rule and thus makes planning possible. So a bad rule may be better than none. A doctor may disapprove of black market transactions but if he can ensure regular supplies of a drug essential to his patients only on the black market this may be better than having no reliable source of supply. A business man may be able to plan his affairs better in a country where every official has his price and the prices are well known than in one where nobody knows whether a service will be rendered without a bribe, and he may take up his time insulting the honest or failing to bribe the venal.

This point about confidence explains why G. E. Moore maintained that a moral rule should be observed only if it is both generally useful *and generally practised*. 'In a society in which certain kinds of theft are the common rule the utility of abstinence from such thefts on the part of a single individual is exceedingly doubtful even though the common rule is a bad one.'[1] Moore admits that this position may be weakened by the possibility that the example given by breaking the bad rule may tend to break down the existing custom and so may have good results. But surely this consideration is *not* how most people would decide the question. It is surely obvious that anyone who is convinced that society would be better off without such thefts (and promiscuity and bootlegging provide similar examples) will have a duty to abstain from them even if they are the established custom and his example will do nothing to weaken the practice. I would therefore reject Moore's conclusion which he states as follows:

> The question whether the general observance of a rule not generally observed would or would not be desirable cannot much affect the question how any individual ought to act; since on the one hand there is a large probability that he will not by any means be able to bring about its general observance, and on the other hand the fact that its general observance would be useful could, in any case, give him no reason to conclude that he himself ought to observe it in the absence of such general observance.[2]

This is entirely in line with Moore's own Utilitarian position by which actual results of individual actions are the test of right and wrong. But I have shown earlier how strong is Ross's argument that some acts are right independently of their actual consequences and I am now suggesting that, at least in some cases, the rule which such acts observe is to be justified by the consideration that if it were generally adopted the results would be good. People act on principle and when acts against their principles are suggested they say 'what would it be like if everyone did that?' I do not see that this argument is weakened in the case where everyone *does* do that—with the results predicted.

1. *Principia Ethica*, p. 164
2. Ibid., p. 161

Moral rules

There is I think one exception here. Where other principles are involved whose application involves following the bad rule I may have a duty to follow it even while I think it bad. I may think tipping is a bad practice and am inclined to refuse to tip (accepting the bad service which results as the price I pay for my principles). But when it is pointed out to me that the fact that the practice is general has resulted in the waiter being paid a minute wage or the car-park attendant no wage at all, I may revise my views.

There is a further argument against the explanation of moral rules as Utilitarian precepts. Problems are bound to arise concerning the *distribution* of good in any society, and these seem to be independent of the amount of good produced and therefore of Utilitarian considerations. And the principles involved for such distribution cannot therefore themselves be regarded as Utilitarian precepts. The simplest of all these principles is that of equality. When there is a budget surplus it might be naturally suggested that a general remission of taxation in equal shares would be fairest. But this might well produce less good results than a large distribution to the neediest group. And this raises another principle, 'To each according to his need'. But it is felt to be unfair to reward the idle (though not the incapacitated). So a third principle would distribute in accordance with work done. It may be possible (and Mill tried to achieve this) to show that principles of justice have in fact a Utilitarian basis—as, for example, that reward for work is needed as an incentive. But this is a very difficult and little worked field and it must be left here as one well worth further consideration.

REFERENCES

On rules as Utilitarian precepts:
    H. W. B. Joseph, *Some Problems in Ethics*, ch. viii
    J. Rawls, 'Two Concepts of Rules', *Philosophical Review* (1955)

J. Harrison, 'Utilitarianism, Universalisation and Our Duty to be Just', *Proceedings of the Aristotelian Society* (1952–3)

Criticism of this view:

J. J. C. Smart, 'Extreme and Restricted Utilitarianism', *Philosophical Quarterly* (1956)

On the problems of justice and distribution:

J. S. Mill, *Utilitarianism*, ch. 5

J. Rawls, 'Justice as Fairness', *Philosophical Review* (1958)

S. I. Benn and R. S. Peters, *Social Principles and the Democratic State*, chs. 5, 6

5

RIGHT, OUGHT, AND DUTY

Moore and Ross, as we have seen, differed greatly about what makes an act right. But they were completely agreed on one central issue. They both thought that, in any given circumstances, there is one act which is the right act for a man to do and this act is his duty and is what he ought to do. What makes it right are the existing circumstances; and the man's own beliefs and motives have nothing to do with it. As we saw, Moore held that motives will come into our judgment of whether an agent was praiseworthy or blameworthy, but not into the question whether he did his duty. There are various corollaries of this view.

(a) My duty does not depend on any belief of mine nor need I know that it is my duty in order to do it. This may be illustrated by two examples, one using Moore's account of what makes an act right and the other using Ross's. On Moore's view that act is right which, of all courses open to the agent, will produce the best results. Now suppose a surgeon diagnoses a disease for which he believes an operation essential but (unknown to him and unknowable to him) the patient has a blood condition which will make the operation fatal. The surgeon's duty here is *not* to operate. With Ross's account, if I am in funds and my parents are in need I ought to help them.

Now suppose they are in need but owing to war conditions we have been separated and I cannot know they are needy. Yet my duty is to help them.

(b) My duty does not depend on my motives. It does not matter, for Ross and Moore, *why* I do the right thing; so long as I do it, I shall have done my duty and done what I ought to do. So a man who pays his debts not because of any sense of obligation but because he fears legal proceedings has done his duty. A man who tells the truth in order to hurt his hearer's feelings (but who fails to hurt him) has done his duty. A man who leaves money to a good cause *solely* in order to spite his relatives has done what he ought.

(c) My doing of duty must be successful. It is no use merely trying. Again the surgeon is a good example. If there was anything he could have done which would have cured his patient, he fails in the duty and does what he ought not to do, if he does not do this curative action. And it is no defence that he tried to cure him or that he did his best.

An effective attack on this view was launched by H. A. Prichard. He maintained that my duty depends on my *beliefs* about the facts and not (as Moore and Ross thought) on the facts themselves. Two kinds of facts are relevant here, facts about the situation and facts about the consequences of actions. So if a surgeon believes a man has smallpox and if he believes that a heavy inoculation will help him it is his duty to inoculate (even if it turns out later that the man had cowpox or that an inoculation would make him worse). If he does this he has done what he ought to do. Prichard goes on to point out that, on the view he is attacking, there are no duties of insurance or precaution. The question whether I ought to stop at a blind cross roads depends on whether there is in fact any traffic coming. I have sometimes been a passenger in a car when the driver took a shocking risk—overtaking over the rise of a hill—and when I showed signs of anxiety he would say, when the danger was over, 'There, you see, it was all right.' For Ross and Moore it *was* all right since in fact there was no danger. But Prichard would say he ought not to have done it; his duty was to hold back until he could see the road was clear.

Prichard's second criticism is to attack the view that my duty is to succeed and not merely to try. Prichard points out that any change in the world which I bring about is the end of a chain of causes which begins with a motion of my body; and once I have performed this motion other causes may come in unexpectedly which will prevent the end result from occurring. Moreover, says Prichard, even the motion of my body is not wholly within my control. Just as I may carve my patient skilfully but peritonitis may set in and he dies; so I may *try* to carve my patient skilfully but fail. All that can be demanded of a man is that he should '*set himself*' to achieve what he *believes* is appropriate to the situation *as it appears to him*. If he does this he has done his duty and done what he ought.

Ross forestalls this second point by an example. Suppose I borrow a book and then pack and post it carefully, but it fails to reach its owner. I have set myself to do what I believe is appropriate. But, says Ross, when I hear it has not arrived, surely I must send him another copy and go on doing this until one does arrive. For one's duty is to return borrowed articles not merely to try to do so. Now I think most people would agree that, when they heard the first copy had not arrived, they would not wash their hands of the matter. But suppose they do what Ross would require. It is surely clear that they are not carrying out the original obligation until they succeed. For in the end the book which gets through is not the 'borrowed article' but a substitute (as would be clear if the borrowed book had the owner's notes or a presentation inscription in it). And the fact that we are not continuing to perform the *same* duty would also be obvious if our reaction, while not one of inaction, was not that of Ross. In such circumstances I think I would go first to the Post Office and see whether they would compensate the lender. If not, then I doubt whether I would simply send him another copy of the same book. I might ask him what he would like or send him a 'book token'. And this makes clear that when I packed and posted the book I *had* done the duty appropriate for borrowed articles. I had set myself to return it (nor, as Ross would say, had I failed in *this* duty). Then when things went wrong I had another different duty, a duty of compensation. This situation is really quite common. A man does what he ought; then things go wrong, and he has another new duty. If it is asked why *he* should have the duty of compensation when the loss is not his fault, the first answer is that he should *not* bear it, the Post Office should. The second is that, if the Post Office will not pay and he does not, then nobody will; and it is more unfair that the lender should end up worse off than that the borrower should. So if I press the claims of a man for a job (in all honesty and sincerity) and he turns out a failure then I have a particular duty to help if I can, not because I failed in my duty (as Ross would say) but because, if I had not pressed his case, he would not have been appointed. So Ross fails to shake Prichard's case that our duty is not to succeed but to try.

Now how are we to decide between Ross and Prichard? In his later book, *Foundations of Ethics*, Ross recognises the strength of Prichard's case but it does not lead him to give up his view of the meanings of 'ought' and 'duty'. He says that it shows that 'ought' and 'duty' are used in two different ways— an 'objective' way defined by success in the actual situation and a 'subjective' way defined by beliefs and efforts. And Ross seems to me clearly correct in this claim.

For objective uses of 'ought' and 'duty' we may quote 'He tried his best to do his duty,' or 'He ought not to have operated but he could not have known that,' where, for the subjective view, to try one's best *is* to do one's duty and where his duty *was* to operate *because* he could not have known that the operation would fail. For examples of the subjective view compare 'Captain Oates did his duty though his sacrifice was in vain'; 'The juryman did his duty in voting "guilty", though in fact later evidence showed the man was innocent.'

What are the alternatives to Ross's conclusion that 'ought' and 'duty' are used both in his way and in Prichard's, and that there is no more to be said? There are only two. One is to give up these ordinary (and ambiguous) words and coin technical terms and give them exact and unambiguous meanings—'deontic' and 'rectal'. The other is to *legislate* for the ordinary words—to say 'duty' and 'ought' are used ambiguously but they *should* be used in Prichard's way only (or in Ross's). Just as whales are sometimes called 'fish' but they should not be.

I started this book by saying that it is concerned to analyse the use of moral words and, if so, the solution in *Foundations of Ethics* would seem the proper one. If words are ambiguous, say so, clearly distinguishing the meanings. Take the word 'pupil'. It would be absurd to say that the word in English means both a student and a part of the eye, but that the latter is wrong and illegitimate.

I am not altogether satisfied by this. If I accept it, I want to add —yes, but the Prichard or subjective use of 'ought' and 'duty' is more accurate, less misleading and dangerous. The Ross or objective view leads to a number of paradoxes. These in fact are no more than striking reiterations of Prichard's criticisms.

First, on the objective view, I can do my duty and do what I ought from a bad motive and with immoral intent. I borrow a book, I fully intend never to return it, I leave it in the lender's room in a fit of absence of mind; I have done my duty, the lender has got his book back.

Second, if I do my duty it will often be by luck. In a review of Ford Madox Ford's *Autobiography* I came across this:

But nothing in the author's brilliant and high-spirited narrative is more striking than his account of how he met in London a lady who said he had changed the whole of her life and saved her from total disaster. Ford said he had never seen her before. 'Yes you have,' she said. 'Ten years ago you came with Willa Cather to visit me in Bronx Park. I was just about to elope with a married man but you stayed so long and talked so much that I missed my train.' Mr Ford says elsewhere that he 'rather dislikes virtue' but he does not say whether he chalks this up to himself as one of his good deeds.

So the inscription on a tombstone 'He always did his duty' could mean 'He was a man who went through life doing what the circumstances required but always either by accident or under compulsion or from a bad motive.' It is all very well for Moore to say that motives have nothing to do with 'right', 'ought', and 'duty' but much to do with moral praise or blame. This is precisely the paradox. We find it very odd to say that the statement that a man has done or failed to do his duty (or done or not done what he ought to have done) has nothing to do with praise or blame. (It is interesting that 'right' and wrong' seem much less paradoxical, as we shall see later.)

So it is tempting to legislate and say that 'ought' and 'duty' are *wrongly* used in Ross's objective sense, or at least to say 'duty' is used in both ways but with a warning about its objective uses.

But, if so, why is the objective use current? There are several reasons. First the subjective view is a comparatively recent and highly civilised achievement. People used to be held to blame for acts for which we now absolve them from responsibility—heretics or lunatics for example. Our recognition in this country that a conscientious objector should not be penalised or victimised or treated as a coward came about only after the First World War. This change seems to me to be an advance, but the older view hangs on. Then again, in earlier times still, unintended crimes and even inanimate objects were punished, in the spirit in which children kick a table which has tripped them up.

A second reason why the objective view is current is that there is a sense of the word 'duty' or 'duties' which is certainly objective: 'sentry duty', 'duty lists', etc. The duties of an orderly officer or of a clerk of court are objective because they are laid down in official documents. They do not depend on my beliefs or anyone's beliefs or motives or efforts.

A third reason why the objective use is current is that if we use 'duty' and 'ought' subjectively we seem still to need *some* objective term. In order to see this let us follow Prichard's argument to its conclusion. His final formula is that I ought 'to set myself to do what I believe will bring about a certain state of affairs'. But this is not enough. The blackmailer or thief sets himself to bring about a certain state of affairs. What is the difference between him and the moral man? It must lie in the state of affairs. We have to distinguish what the good man sets himself to bring about from what the bad man does and for this we need a moral term at the end of our formula. The term 'right' would serve. A man does his duty (and is a moral man) if he sets himself to do what he believes to be right. The bad man sets himself to do what he believes to be wrong. Here 'right' and 'wrong' are used objectively, and this would be another reason for legislating about the use of 'duty' and 'ought'. For if 'right' is used objectively, 'duty' and 'ought' so used would simply be synonymous with it.

But there is a further element of subjectivity to be considered. Prichard had argued that my duty depends on my beliefs about facts. But, as E. F. Carritt pointed out, there are other beliefs which have to be taken into account—beliefs about moral principles and moral values. Sometimes this is not obvious, because moral agreement is taken for granted. A surgeon's duty depends on his factual beliefs; if he believes an operation will save his patient's life his duty is to operate. Another surgeon who believed an operation would kill his patient would have a duty not to operate. Their different duties depend on their different factual beliefs. We take it for granted that the moral principle involved—that life should be saved—is one on which they are agreed. But this is not always the case. One doctor may recommend a heavy dose of morphia —the other says 'that will kill him'—the first replies 'Yes, I know.' Here they agree on facts and disagree on moral standards. One man may reject another's bellicose reaction because he believes in the efficacy of passive resistance—here the disagreement is factual; but he may reject it because he believes violence is never justified—this is a moral conflict. Carritt accepts all the subjectivity in Prichard but goes further. On his view a man's duty is to set himself to do what his own moral estimate of the supposed situation requires. ('Set himself' and 'supposed' are Prichard's subjectivisms; 'moral estimate' is Carritt's addition.) There is a special difficulty in connection with moral beliefs. We might be inclined to agree that the duty of a pacifist is to refuse to fight in any cause however 'just'. But there are extreme cases where we may feel less certain. What about the fanatics: Nazis persecuting Jews, inquisitors torturing heretics, communists brainwashing prisoners—all believing these activities are right and proper? Are they to be regarded as doing their duty and earning moral commendation? What is the difference between them and the pacifist? I am inclined to think that there is no difference, and that we must accept the paradox and award Hitler moral approval. Why are we less inclined to do so than in the case of the pacifist? It may be because we are not so absolutely certain that the pacifist is wrong; it may also be because the pacifist is overemphasising a value we accept—the evil of war and the merits of peaceful and persuasive solutions; it may be again because pacifists (Quakers for example) are in all other ways people we admire. But it has to be remembered that toleration of pacifists is a very recent phenomenon. In the First World War they were prosecuted and otherwise persecuted with very general popular support. So we may be in a transitional stage in this matter and tolerance of people we now consider fanatical may well come. It should also be remembered that moral praise is only one kind of valuation. One may disapprove of the fanatic in other ways, and one may even (without blaming him) have a duty to restrict him (as one would a lunatic) or destroy him (as one would a mad dog).

There is one final added element of subjectivity to be considered—the most doubtful and debated of all. That is motive. It may be said that a man might satisfy Carritt's formula and still not be said to do his duty. For example, I am an elector to a Chair of philosophy; it falls vacant and I have to give my vote. Have I done my duty? What moral principle do I believe relevant? I believe the best man should be elected. What relevant beliefs of fact did I entertain? I believed Jones was the best man. Did I set myself to get Jones elected? I did; I voted for him. So I did my duty. But suppose my motive was friendship for Jones or malice against his rival. Then I should have voted for Jones whether I believed him the best man or not. I said to myself: 'What a bit of luck that his record is the best too. No one can cast a stone at me.' (We recall how, on the objective theory, the doing of duty is often a matter of luck.) The result of refusing to include motive in the definition of duty is the same paradox as we noted before, the separation of praise and blame from concepts of 'duty' and 'ought'. For obviously I deserve no praise for voting for my friend or voting from dislike of his rival.

Ross and Prichard both provide arguments against including a reference to motive in the definition of duty. First, it is said, such inclusion results in a regress; for what is the motive whose presence would be required? Surely a sense of duty.

My duty will then be 'to set myself to do what I believe appropriate to the situation from a sense that it is my duty'. That is 'from a sense that I ought to set myself... from a sense of duty'. But the regress arises only if we identify, as Ross does, duty with right and hold an objective theory of duty. If, as suggested above, we distinguish 'duty' from 'right' and use 'right' objectively, our formula now becomes 'My duty is to set myself to do what I believe to be right because I believe it to be right.' The regress or circle vanishes.

The second argument against including motive in duty is that 'ought implies can' and that our motives are not within our control. 'Ought implies can' is a central and agreed principle among all the moralists whose views we are considering. There are great problems about its analysis, but for the moment we must leave them aside. What is being maintained is that when I say someone ought to do something this implies that he can do it. Now the objection is that I cannot have a motive at will. I can decide to get up or to help a friend. But I can't decide to be angry or hungry, to love or to hate. Yet these are typical motives. There are two answers to this argument; the first goes a long way to meet it; the second, I think, goes all the way. First our motives are not wholly independent of our will. Ross admits that we can *cultivate* a motive. But this is not enough because it is a long-term plan and the definition which includes motive would seem to require us to have the appropriate motive *now*. Ross would still deny that we can alter our immediate motives by immediate choices. But sometimes we can. When people are angry we tell them to count ten. This reduces their anger. But this is to alter an exisiting motive not to produce a new one. But I can produce anger by recalling an injury someone has done me. I can generate hunger by imagining Lucullan feasts, or love for Chloe by dwelling on her charms. But it may be said that these procedures are indirect and consequently may fail. I may find I can't be angry after all and no pictures of feasts can rouse my appetite. And after all there may not be techniques known to me for arousing any motive. So this is a very limited answer.

But it leads to the second answer which seems entirely satisfactory. The motive required is the sense of duty. Now what is the technique for having this motive? It is simply to consider the rights and wrongs of any situation. It makes sense to say 'I can't be angry with Jones this morning; I can't be angry with anyone; the sun is shining and I'm brimful of happiness' or 'I can't feel hunger this morning. I have tried every device but I can't.' But it would sound very odd if a man said 'I find I can't have a sense of duty this morning.' This is because he would be saying that he could not recognise anything as right or wrong, good or bad.

We may recall Saki's Clovis:

'My mother never bothered about bringing me up. She saw to it that I got whacked at decent intervals and was taught the difference between right and wrong; there is some difference, you know, but I've forgotten what it is.'

'Forgotten the difference between right and wrong!' exclaimed Mrs Eggelby.

'Well, you see, I took up natural history and a lot of other subjects at the same time and one can't remember everything, can one? I used to know the difference between the Sardinian dormouse and the ordinary kind, and whether the wryneck arrives at our shores earlier than the cuckoo, or the other way round and how long the walrus takes to grow to maturity. I daresay you knew all these things once, but you've forgotten them.'

'These things are not important,' said Mrs Eggleby, 'but . . .'

'The fact that we have both forgotten them proves that they *are* important,' said Clovis, 'you must have noticed that it is always the important things one forgets while the trivial unnecessary facts of life stick in one's memory.'[1]

1. *Beasts and Super-Beasts* (collected 1926 ed). p. 159

It may well be that he cannot see anything wrong in some particular action and therefore has no sense of duty about *it*. But (on the subjective theory we have so far accepted) he would not then believe it to be wrong and would have no duty to abstain from it.

There may indeed be exceptional situations in which a man is unable to see *any* distinctions between right and wrong. It may be that extremes of torture or exhaustion or starvation, or the use of certain drugs or of Russian methods of brainwashing or (as some claim) the brain operation of frontal lobotomy—one of these may make recognition of all or any moral distinctions impossible. Then a man can't have the motive of duty. But under these circumstances he can't have duties either. It makes no sense to say that he ought not to have betrayed his colleagues or confessed to sabotage or treachery. So the exception confirms the rule. Unless the motive can be present duty cannot be present either.

There is another way in which this conclusion can be reached. We started from the difficulty that I might set myself to do what I believe to be right from various motives. My motive in voting for Jones might be duty or friendship or malice towards Jones's rival. But this assumes that I can set myself to do the same action from different motives. But is this really so? An action is not just a bodily motion—an affixing of an ink X to a form—it is a psychological affair. What am I really setting myself to achieve? Surely in one case to get the best man elected (or Jones qua best man) in another to get my friend elected (or Jones qua my friend) and in the third to do down Robinson by voting for Jones. Suppose that I set myself to get Jones elected qua best man but that I knew in addition that Jones was a Freemason or a three-handicap golfer. Could my enemies say that I had set myself to get a Freemason or a golf-tiger elected? So when I set myself to get my friend elected knowing him to be the best man it is equally inappropriate to say that I set myself to get the best man elected. Any distinction between bodily movement and willed action, and any distinction between involuntary and voluntary action, has to include a reference to motive as a part of willed voluntary action. So what happened in this election case was not that I set myself to do the right thing from the wrong motive, but that I did not set myself to do the right thing at all. I set myself to do an act of favouritism or spite which by good fortune had the same external character and actual results as the act I would have done if I had set myself to get the best man elected.

To sum up, Moore and Ross said duty was objective. To know whether a man had done his duty or done what he ought, one need know only the facts about the situation and the actual consequences of his action. One need know nothing at all about his beliefs and motives. I have argued, following Prichard (on factual beliefs), Carritt (on moral beliefs), and my own nose (on motives), that in order to know whether a man has done his duty one must know:

(a) his beliefs about the situation (not the facts about it)
(b) his beliefs about the consequences of his action (not the actual consequences)
(c) his moral beliefs
(d) his motive
(e) what he set himself to do (not what he actually achieved).

REFERENCES

W. D. Ross, *The Right and the Good*, pp. 4-7, 31-2, 42-7; *Foundations of Ethics*, chs. vi, vii

H. A. Prichard, 'Duty and Ignorance of Fact', *Proceedings of the British Academy* (1932), reprinted in *Moral Obligation*

E. F. Carritt, *Ethical and Political Thinking*, ch. ii

For defences of the objective theory of duty:

    H. Nystedt, 'The Problem of Duty and Knowledge', *Philosophy* (1951)

    K. Baier, 'Doing my Duty', *Philosophy* (1951)

For the subjective side:

    H. J. N. Horsburgh, 'Baier on Doing one's Duty', *Philosophy* (1952)

    N. H. G. Robinson, 'The Moral Situation', *Philosophy* (1949)

On the relevance of motive to duty:

    I. Gallie, 'Oxford Moralists', *Philosophy* (1932)

    G. E. Hughes, 'Motive and Duty', *Mind* (1944)

    H. D. Lewis, 'Moral Freedom in Recent Ethics', *Proceedings of the Aristotelian Society* (1946-7)

On motives and 'ought implies can':

    J. Wheatley, *Analysis* (January 1962)

    A. Savile, *Analysis* (March 1963)

    J. Wheatley, *Analysis* (December 1963)

    P. Taylor, *Analysis* (October 1964)

6

THE DUTY TO THINK

Throughout the last chapter it has been repeatedly urged, in favour of all the subjective criteria for 'duty' and 'ought', that only with these criteria can praise be associated with 'ought' and blame with 'ought not'. So a man is to be praised if he sets himself to do what he believes to be right because he believes it right. But there seem to be exceptions to this. We use 'well-meaning' as a term of blame and we say that the path to Hell is paved with good intentions. When do we do this? When a conscientious action leads to bad results. But not in all cases. We should not describe an eighteenth-century surgeon who operated without chloroform or a nineteenth-century physician who failed to prescribe penicillin as well-meaning. This is because they could not have known any better. We condemn the well-meaning man in cases when he could have found out the relevant facts or thought more effectively about them.

Here, however, we are blaming him not for a present dereliction of duty but for a past one, as the tenses of our verbs show. 'Why did you do that?' 'I thought it would help him.' 'But don't you know he can't bear that sort of help?' 'Yes, of course, but I didn't think of it.' 'Well, you *ought to have thought*.' But it follows that the man is still to be commended for doing what he does *at the time*. For he can *then* do no better. He is making the best of a bad job. A single-handed village doctor takes a glass too much sherry at a party and then is called out to an accident. It is no good saying his duty is to drive his car as he would if he were sober; he can't. If he drives it as carefully as he can (in his actual condition) he has done his duty. He can do no more. He ought not to *have had* the extra sherry. This case is clearly parallel to those in which failure to think precedes by a long interval the time of action. Suppose I am asked to read a paper in Scotland early in October, and I do not stop to consider that this is the beginning of the Oxford term and accept the invitation. When the time comes I must make the best of a bad job, either cancel my paper or absent myself from Oxford at a time when a great many people want to see me. Whichever of these I do will be the best of a bad job. I ought never to *have* accepted. No, but October is here and I have a duty now; which will be very different from the duty I would have had if I had not failed in the duty of thinking.

People may find it very hard to commend the well-meaning action at the time it is done. But if you ask them what else they would have the agent do (in the circumstances in which he has landed himself), they will see the reason for commending it. One of the difficulties they meet is that the duty of thinking may be so close in time to the duty of action that it is difficult to distinguish them.

An extreme case of this is raised by Professor Ryle's account of intelligent action. He argues that in many cases (perhaps in most, and certainly in the best cases) there are not two distinguishable processes, thinking and acting. A really intelligent games player does not think first and act afterwards, nor does an intelligent speaker. Intelligence is shown not by a prior process of thinking but by the way the act is done—intelligently. And even when plans are laid by prior thinking, they

The duty to think

have still to be carried out intelligently. Now what happens to my solution of the 'well-meaning' problem? I think I must say that there is a duty not only to act but to act intelligently (thinking what you are doing). And the well-meaning man while he satisfies the other factors in duty fails to satisfy this one. There are parallels. I may have a duty to tell a man an unpleasant truth and this includes a duty to tell him it *tactfully*, and in this respect I may fail. Or I may have to announce a decision and to announce it *firmly* (so that there is no doubt about its finality) and in this respect I may fail. Perhaps we would not call a man who fails to perform a duty thinkingly 'well-meaning' or 'well-intentioned'; we should more likely call him 'casual' or 'careless' or 'slapdash'—but these would still be words of blame. Perhaps 'negligent' covers both cases. We tend to associate 'intention' with prior thinking (again mistakenly as Ryle would urge; much is intended which is not precognised).

The second difficulty about this explanation of 'well-meaning' is that the duty to think appears unlimited. Prichard says that, before action, I must 'consider as fully as I can' whether my action will be appropriate.[1] But at first sight this would seem to have obvious dangers:

> The native hue of resolution
> Is sicklied o'er with the pale cast of thought.
> And enterprises of great pitch and moment
> With this regard their currents turn awry
> And lose the name of action.

Hamlet's trouble was 'thinking too precisely on the event'. It might be said that Prichard meant by 'as fully as I can' 'as fully as the circumstances allow' and this would take care of the danger of missing the bus. But there are other cases where excessive thought would be condemned, where much thought

1. *Moral Obligation*, p. 27

is given to a trivial decision and Prichard could not cover this case. So now it is tempting to say that the amount of thinking I have a duty to do is limited by the time available and by the importance of the decision.

But once again the subjective factors return upon us. How can a man be sure how much time is available? I once saw a man go through the ice on a lake, and taking some risks I went over cracking ice for a ladder. If I had stopped to watch him I would have seen that he was sinking quite slowly into mud, so I had much more time available than I thought. But I acted on the belief that time was short. Or again can we always be sure what decisions are important? The decision which has most greatly affected my life seemed at the time quite trivial—whether to make a social call rather late at night on my pre-war professor. The call changed my whole career. So here too we must devote as much time as *we believe* to be available, or as much as the importance which *we attach* to the decision seems to require. So what began as a difficulty in the subjective theory of duty turns out to be another duty subjectively qualified.

It follows that the thinking may be unsuccessful and the results still unsatisfactory. In particular we can take account only of considerations which 'occur' to us; and, as 'occur' indicates, this is something over which we have no direct control. We may cultivate interests and acquire knowledge to help the right ideas to occur; and we can take counsel and advice so that ideas which may occur to other people may often be recognised at once as relevant to our problem.

There is however a final difficulty about the 'duty to think' as a solution of the use of 'well-meaning'. The 'duty to think' seems to be an *objective* duty. For it is no defence to say 'I did not believe I had to think, it never occurred to me to stop and think'. To fulfil the duty of thinking is *not* to set myself to do what I believe to be right. This difficulty can be partially met by pointing out that it is impossible to do any duty without *some* thought; so what is complained of is that I didn't think enough. But 'enough' brings us back to the questions raised earlier about the limits on the duty of thinking. Nevertheless it does seem that there is a genuine problem here.

REFERENCE

Ryle on acting intelligently:
 The Concept of Mind, ch. ii

7

VARIATIONS IN MORAL BELIEF

Up to this point our argument has followed the example set by Moore, Ross, Prichard, and Carritt in appealing, in support of any moral theory, to our actual use of moral words. How could you decide...? What would you say if...? Moore's argument that pleasure is not the only thing good in itself, good as an end; Ross's demonstration that production of the maximum good is not the sole duty of man; Prichard's subjective theory; Carritt's addition to it;—all these rest on such an appeal. But even if these arguments succeed all they show is that these theories fit the moral discourse of ourselves and of these philosophers. Some people do not agree with us and with them. Some would hold against Moore that knowledge is not good in itself, or that aesthetic enjoyment is good only for the pleasure it involves (and Puritans might reject even that). Others would hold against Ross that lying is not wrong as such; and, as we have seen, it is only recently that we have come to accept the Prichard-Carritt view of the conscientious objector. Most people in 1914 held that he did not do his duty. So the theories mirror the morals of mid-twentieth-century middle-class Britons. And a theory based on the use of moral terms in China or Peru might look very different. Views on morals differ and it is to be expected that the theories built on them will differ too.

Variations in moral belief

In other fields, when beliefs differ we normally appeal to argument and evidence to help us to decide which view is correct. And indeed the final conclusion of Chapter 5 that a man's duty depends on his *moral* beliefs as well as on his factual beliefs implicitly recognised both the differences in moral beliefs and the consequence that when beliefs differ one must be mistaken. The same point was implied by our acceptance of 'right' as an objective term, though 'duty' remained subjective (p. 57). For how are we to decide what *is* right when beliefs differ?

As we have seen, most actual moral judgments are a fusion of factual and moral beliefs. 'I ought to pay Jones the money' rests on 'I borrowed the money and have not yet repaid it' (which is a factual belief) and 'debts should be repaid' (which is a moral belief). The factual beliefs can be verified or falsified by ordinary argument and evidence. But what about the moral beliefs? We have seen in Chapter 4 that some philosophers (Ross and Prichard) thought moral rules required no validation because they were self-evidently true, but that this line and the attempts to defend it are unconvincing. In that chapter it was argued that a moral rule is justified, if the consequences of its general acceptance would be good. This justification again breaks up into two elements, the factual element (what will the consequences be?) and the moral element (would these consequences be good?)

Now some ordinary men, and some philosophers too, would reject this Utilitarian derivation of moral rules. A pacifist would reject violence as a general method of solving problems, no matter how good the consequences were; and so too with certain churches in regard to divorce or birth-control. But even if this derivation is accepted we are still not able to provide a complete verification procedure for moral beliefs. For the question remains 'which results are good?' And, as we have seen, some have rejected the view that knowledge is good or that beauty is good or that pleasure is good. Moore's appeal here is to self-evidence and it is as inconclusive as that of Ross on moral rules. But here indeed there seems to be a deadlock. For what types of arguments could be involved?

When a statement is attacked we sometimes defend it by showing that it follows from some other statement deductively. Why is the hydrogen bomb a bad thing? Because anything involving the mass destruction of innocent people is bad. But this only pushes the question back. And clearly to prove in this way that anything was good in itself or good as an end would seem self-contradictory.

Induction or an appeal to facts would seem equally suspect. For the facts here are the moral judgments of men and we have seen that they differ and contradict each other.

In both cases the difficulty is one put very clearly by Hume.

> In every system of morality, which I have hitherto met with I have always remark'd, that the author proceeds for some time in the ordinary way of reasoning, and establishes the being of a God or makes observations concerning human affairs; when of a sudden I am surpriz'd to find that, instead of the usual copulation of propositions, *is* and *is not*, I meet with no proposition that is not connected with an *ought* or an *ought not*. This change is imperceptible; but it is, however, of the last consequence. For as this *ought*, or *ought not*, expresses some new relation or affirmation, 'tis necessary that it shou'd be observ'd and explain'd; and at this same time that a reason should be given, for what seems altogether inconceivable, how this new relation can be a deduction from others which are entirely different from it.[1]

Conclusions containing the notion of ought (or ought not) cannot be derived from premises none of which contain such notions. Or, more generally, to include good and bad also, conclusions involving any moral terms cannot be based on

1. *Treatise of Human Nature*, book III, part I, section 1

premises none of which contain any moral terms. Thus it would seem that standard methods of argument *must* fail to validate or falsify moral judgments. This important principle has been called (by Professor Max Black) 'Hume's guillotine'.

There is, however, another kind of defence which might seem to escape Hume's guillotine. When a statement is attacked it is sometimes defended by showing that it is supported by a lot of other similar statements—this is sometimes called the 'coherence test' and is applied to the stories of witnesses in cases where external verification is not obtainable. But there are difficulties here too. First, the moral beliefs of a man are seldom coherent in this sense. Indeed as we have seen they often clash with each other. But, even if they were coherent, coherence cannot guarantee truth, though incoherence does indicate falsehood. A witness's coherent story may be completely false; or a barrister's coherent reconstruction of a crime may be completely false as based on the evidence of perjured or collusive witnesses. Some moral principles do cohere (as we shall see later) in such a way that altering one involves altering others and so that they together form what has been called a 'culture pattern'. But there may be rival culture patterns and how is one to be preferred to another? 'Coherence' cannot tell us that.

REFERENCES

Hume's guillotine

Attacked: P. Foot, *Proceedings of the Aristotelian Society* (1958–9)
J. R. Searle, *Philosophical Review* (1964)
M. Black, *Philosophical Review* (1964)
G. I. Mavrodes, *Analysis* (December 1964)

Defended: A. Flew, J. E. McLellan and B. Komisar, D. Phillips, all in *Analysis* (December 1964)

8

THE EMOTIVE THEORY

The subjective theory to which the argument of Chapter 5 led was a subjective theory of *duty* and was combined with an objective theory of *right*. The formula was 'I must set myself to do what I believe to be right'. It was pointed out that my belief that a particular action would be right is always the resultant of certain *factual* beliefs about the situation and the probable consequences of the action and of certain *moral* beliefs about what in general is right or good. In the previous chapter we have seen that there are difficulties about verifying moral beliefs when these differ. This difficulty has led philosophers to ask whether the moral element in an assertion of right or wrong is really a *belief* at all. For, if it were a belief, it would have to be true or false; and there seems no way of showing whether it is the one or the other. What makes a belief true or false must be the facts. For a moral belief, it must be the moral facts. But these moral facts are inaccessible. So let us deny there are any moral facts. I say that divorce is sometimes right; you say it is always wrong. I say that Jones's doctor ought to tell him the truth about his disease; you say he ought not. We find we cannot establish whose belief is true: that is, we cannot determine the moral facts about divorce or truth-telling. So there are no such facts. So what we called our 'beliefs' cannot be beliefs. What then are they?

What do the words 'divorce is wrong' express, if not a belief? Hobbes had a first shot at this. When a man says 'this is good' he means to state that he desires it.

> Whatsoever is the object of any man's Appetite or Desire; that is it, which he for his part calleth *Good*: and the object of his Hate and Aversion, *Evill*; and of his Contempt, *Vile* and *Inconsiderable*. For these words of Good, Evill, and Contemptible are ever used with relation to the person that useth them: There being nothing simply and absolutely so; nor any common Rule of Good and Evill to be taken from the nature of the objects themselves.[1]

So when a man says 'truth is good' he is speaking not about truth but about himself and saying he desires truth. When he says 'Communism is bad' he is not describing or saying something about Communism, for Communism is not itself either bad or good. He is describing his own aversion to it. Now this theory, though very interesting as such an early specimen, will not do. It cannot explain the *moral* use of 'good' or 'right', because there is often a conflict between duty and desire; and then I am compelled to say 'It is not good but I want it'— (drugs, for example, or my rival's death). But for Hobbes this would mean 'I do not desire it but I do desire it.'

A theory which meets this difficulty was that of Hume, by which to say something is 'good' is to say that I approve of it, and 'bad' that I disapprove of it.

> Take any action allowed to be vicious: wilful murder, for instance. Examine it in all lights and see if you can find that matter of fact or real existence which you call *vice*. In whichever way you take it, you find only certain passions, motives, volitions, and thoughts. There is no other matter of fact in the case. The vice entirely escapes you as long as you consider the object. You can

---

1. *Leviathan* (Everyman edition), ch. 6, p. 24

never find it till you turn your reflexion into your own breast and find a sentiment of disapprobation, which arises in you, towards this action. Here is a matter of fact. It lies in yourself, not in the object.[1]

So the formula with which this chapter opened must lose its implication of objective rightness by losing its reference to *moral beliefs*. It will now run 'My duty is to set myself to do what I believe will bring about *what I morally approve* in the situation as I believe it to be.' Here the factual beliefs (about the situation and the results of my action) are retained because they can be verified and the facts made known.

In this theory, good, bad, right and wrong function like 'nice' and 'nasty'. If I say 'beer is nice' this means I like it. If you say 'it is nasty' this means that you dislike it; and both these propositions are true. I like beer and you dislike it. 'Beer is nice' 'beer is nasty' seem to be about beer and seem to contradict each other. But they are about you and me and are quite compatible with each other. It would of course be a contradiction for *me* to say 'beer is nice' and 'beer is nasty' because this would mean that I both like and dislike it. In most cases the truth of a proposition is independent of the person of the speaker. If I say 'Queen Anne is dead' this is just as true if you say it. But in some sentences the truth depends on the speaker; for instance 'I am short-sighted' which is true if said by me and false if said by you. This is because the 'I' refers to the speaker. Now, on the Hume theory the truth of the statement 'beer is nice' depends on who says it, though it does not look as if it did because it has no reference to 'I' in the actual words. It is misleading in this respect. Similarly with 'Nero was a bad man' which seems to say something about Nero, as in 'Nero was a stout man'. But it

[1]. *Treatise on Human Nature* (ed. Selby-Bigge), book III, part I, section 1, pp. 468-9

does not; it tells us about the speaker, that he disapproves of Nero. It is to be noted on this theory that moral propositions are still propositions; they are still true and false; there are still moral facts, but they are propositions, truths, facts about the speakers who enunciate them.

A variant of this view has recently been put forward by, among others, Professor A. J. Ayer. For him moral judgments are not propositions at all; they are not true or false; they do not *describe* anything, not even the feelings of the speaker. They are more like exclamations; they *evince* or *show* approval and disapproval. Ayer's argument for this is that

> we can reject the view that a man who asserts that a certain action is right or a certain thing is good is saying that he himself approves of it on the ground that a man who confessed that he sometimes approved of what was bad or wrong would not be contradicting himself.[1]

This argument seems unsatisfactory. If the terms 'bad', 'wrong' and 'approved' are all used *morally* then a man could not confess what Ayer suggests. He might morally approve of something which was legally wrong or bad for his purse or his health. Or he might non-morally approve (aesthetically or as an object of desire) of something which he would describe as morally bad or wrong. What he cannot do is to say 'I feel moral approval of this action or thing but it is morally wrong or morally bad.' And indeed Ayer himself a few pages later[2] makes this admission: 'If I say "Tolerance is a virtue" and someone answers "You don't approve of it" he would on the ordinary subjectivist theory be contradicting me.' I should say it is obvious on *any* theory that he could be contradicting me. It may also be that Ayer was misled by his word 'sometimes'. I can say 'Regulus did the right thing though I have

1. *Language, Truth and Logic* (2nd ed.), p. 104
2. Ibid., p. 109

sometimes morally disapproved of his action.' I can certainly say 'This is bad or wrong, though I used to approve of it morally when I was younger.' Or Ayer may mean (by 'sometimes') 'in some exceptional cases'. But if he approved of x in this exceptional case he could not hold that in this case it was wrong. A similar point on which the subjectivist theory must be clear is that the approval which a moral judgment describes is *present* approval though the verb in the sentence concerned is past or future. 'Nero was a bad man,' 'You will do right if you tell him the truth' must be analysed as 'I *now* disapprove of Nero's character though Nero is dead' and 'I *now* approve of your hypothetical future act of truth telling.'

Though Ayer's argument for rejecting the view that moral judgments describe the speaker's feelings may be mistaken, it is still possible that the view itself may be right. A man who says 'That *was* a shocking thing to do' does not seem to be giving a calm and dispassionate account of his own feelings. Indeed 'How shocking of you!' would often be an equivalent utterance. An exclamation is normally taken to be an expression of a feeling involuntarily forced from the utterer of it. A groan evinces pain as sweating evinces embarrassment or pallor evinces fear. The man who in *The Hunting of the Snark* shouted 'Hi' or some other loud cry was simply giving vent to his feelings. But some exclamations may be used to *communicate* my feelings. I am offered a glass of wine to taste and I say 'Lovely'. But this means that I can use an exclamation to deceive. I can say 'Lovely' when I taste ouzo because my host is a Greek and I do not want to offend him. I can say 'Ow' at my dentist's before he hurts me because I think he is getting very near the nerve. It is to be noted that not all exclamations can be so used. If I see a man dancing about I have no way of knowing whether he is rejoicing or furious, or his feet are hurting him or the bricks are hot. So he cannot deceive me. But exclamations which have a conventional attachment to certain feelings like 'Ow' and 'Boo' and 'Lovely' and 'Hurrah' can be used to communicate feelings and therefore to mislead. Ayer says that a moral judgment expresses feeling 'by a suitable convention'. But it thereby communicates the fact that I have the feeling. And this is not so far off from describing the feeling, as in Hume's theory. Indeed it would seem that the best line for a holder of the Emotive theory to take would be to say that moral judgments lie somewhere between pure descriptions and pure exclamations and have some of the characteristics of both.

The fullest working out of the Emotive theory of ethics is to be found in the writings of C. L. Stevenson. He makes a number of useful points, and one important addition to the theory as we have seen it in Hume and Ayer. He distinguishes conflicts of belief from conflicts of attitude. Many disagreements about what we ought to do or about what is good or right are not really moral disagreements at all. One doctor says 'We ought to operate', the other says 'No, we ought to try antibiotics.' They are in complete moral agreement (that the patient should be cured); they differ in belief about what will cure him. Sometimes such a quarrel may therefore be quickly settled. 'He did wrong to leave his wife and children and go off to Peru like that.' 'No he didn't.' 'What? Not wrong to desert his family and leave them unprovided for?' 'He didn't. They are going out to join him on the next boat.' 'Oh, I didn't know that.' Here from the beginning there was moral agreement between the disputants (wife-desertion is wrong); the question was one of fact and of conflicting beliefs about fact. This is Hume's distinction brought out clearly. The point Stevenson makes is that the vast majority of disagreements about what is right or my duty turn out on examination to be disagreements in belief and therefore in principle soluble by evidence. The moral is that, even if we accept Emotivism, it is wrong in ninety-nine cases out of a hundred to apply the theory *simply* in every case. When I say 'x is right' and you say 'x is wrong' it is almost always an error to say this just means that I approve of it and that you disapprove of it; and there is no more to be said, because morals are a matter of taste not argument. No doubt the first move is correct. When I say 'x is right' then I do approve of it. But we must ask 'why?'. I may be afraid of that creature in the field. There is no doubt that I *am* afraid of it. But why? Because I think it is a bull. Then I see it isn't. My fear vanishes. He disapproved of Jones's departure to Peru. Yes he did. But why? Because he thought it was desertion. When he found it was not, his disapproval vanished. For basically my fear is of bulls and his disapproval is of wife-desertion.

Having emphasised this good point, Stevenson then makes a further equally good point. I have suggested that it is normally easy to find out whether a disagreement is one of belief (factual) or one of attitude (moral) by asking 'why?'. But this is not always so easy—as in the case of the poet who did not love Dr Fell. An obvious recent example is Nuclear Disarmament. Some of the arguments are clearly the one or the other. 'If we give up the bomb other states will follow our example.' This is a purely factual prediction, for which there can be evidence, and for which there could in principle be verification. 'If giving up the bomb meant that our country would be overrun without resistance and annexed by the USSR, I don't care. I'd rather die or live under communist rule than be a party to the use of such a weapon.' This is a pure moral preference. There are no facts which could be brought against it (or to support it either). But apart from these extreme cases, the rest of the CND issue is certainly one in which the moral and the factual elements are hard to disentangle.

Stevenson goes on to point out that language itself frequently helps to blur the distinction between moral and factual elements. Many terms appear to be factual but are also morally loaded and they can therefore be used to deceive. 'We can't give him a vote; he's a nigger.' 'We can't give him the job; he's a Yid.' 'We can't trust him; he's a Red.' In each case a reason seems to be given for the actions in question. And the reason seems to state a fact; and to some extent it does. But it also expresses an attitude. In these cases the trick is obvious; no one could be misled, because words like 'nigger', 'Yid', and 'Red' stink. But it is not so easy. What is the non-stinking and purely factual alternative to each of these words, 'Negro', 'Jew', 'Communist'? But these too can function as condemnatory, by tone of voice, by emphasis, and by context. For a pleasant example we may recall A. E. Housman's obituary of Arthur Platt, which ends 'He was addicted to tobacco, he was indifferent to wine, and he would spend long afternoons watching the game of cricket.' Housman prefaces this with the sentence 'I must now enumerate his vices'. But, even without that, 'addicted' and 'indifferent' would give the clue; and, in that context, it would be obvious what Housman's own attitude to cricket was. Stevenson does not note the difficulty of finding non-loaded terms as equivalents. But, in order to see this, let anyone take a strong controversial letter to a newspaper and underline the loaded words. And then let him try to substitute 'aseptic' words for them, words which give the same information but without bias. I once attended an enquiry on a Road Plan. The counsel attacking the road said to the expert witness who had devised it: 'You will agree, I take it, that the object of this is to inject another thousand cars a day into St Giles' Street.' With an eye open for this kind of thing I noticed the word 'inject'. The expert witness could not deny this suggestion. To my great delight the Counsel defending the road arose at once to re-examine. 'You will agree, I am sure, that the traffic which blocks the High Street cannot be relieved unless a large number of cars are led into an alternative route.' 'Led'—'lead kindly light' and so on! Of course the expert had to answer 'yes' to this question too. I said to myself 'What is the aseptic alternative to "inject" and "lead"?' And I found there was none. This is not surprising, because language, like other social institutions, is commonly used to win friends and influence people; so it is to be expected that little of it (except scientific terminology—here perhaps 're-route'!) is aseptic and non-loaded.

The last point on which I think Stevenson makes an advance on Ayer is that he describes the state of mind concerned with moral approval as an 'attitude' rather than as an 'emotion'. There is an admission by Ayer which indeed takes us towards this change. In the second edition Preface to *Language, Truth and Logic* he says[1] that he failed in the first edition

to bring out the point that the common objects of moral approval and disapproval are not particular actions so much as classes of actions ... if an action is labelled right or wrong or good or bad as the case may be it is because it is thought to be an action of a certain type.

But, in that case, the state of mind of the speaker is obviously better described as an 'attitude' than as an 'emotion' or 'feeling'. To say that he judges an action right or wrong because it belongs to a certain class is to say that he would tend to judge other actions belonging to that class right or wrong too, and this is a hypothetical proposition or a group of them. Ayer goes on to say—but in words which show he has not wholly mastered the change from 'feeling' to 'attitude'—

What seems to be an ethical judgment is very often a factual classification of an action as belonging to some class of actions by which a certain moral attitude on the part of the speaker is habitually aroused.

---

1. p. 21

But 'aroused' is ambiguous. It may merely mean that the recognition of the action as belonging to the class provides an occasion for the attitude to be *expressed* (and this is what 'habitually' suggests). But a general tendency cannot be said to be 'aroused' by the occasions for its expression. Squeezing a piece of rubber does not *arouse* its elasticity nor does the writing of this chapter *arouse* my capacity to write grammatical English. Of course 'arouse' might mean that the occasion actually altered my attitude. I am sorry for the Armenians; I have some tendency to pity them when I hear of their sufferings and to subscribe to Armenian Rehabilitation Funds. But when I hear Gladstone and, like his other hearers, I find my pity 'deeply aroused', this means that I tend to do things which, without Gladstone's oratory, I would *not* have tended to do. Tears flow; I send unusually large donations to the Fund; I demand war with Turkey.

Stevenson too is inconsistent about his term 'attitude'. It is true that he explicitly prefers it to 'emotion' and for the right reason. Response is to a range of emotions or rather 'an attitude (which is itself a complicated conjunction of dispositional properties)'.[1] To say I have an attitude to a situation is to say that I am reacting to a certain feature of this situation *and* that I should react similarly to any similar situation. Yet, when Stevenson speaks of moral rules, he traces these to 'psychological economy', to 'the convenience of generalisation'.[2] 'The hearer like the speaker will instinctively avail himself of the psychological economy that comes of ordering the objects of his attitudes in some rough sort of classification.' This suggests that reaction to particular cases could be independent of and prior to their classification. But in that case 'attitude' is not the right term. There are two possible alternatives: (a) I have feelings towards particular actions. These

1. *Ethics and Language*, p. 60
2. Ibid., p. 95

feelings are actually occurrent and not dispositional. I can then notice similarities between the events which arouse these feelings, and make rough classifications of them for convenience; or, without such explicit attention, I can form a habit of having such feelings. Or (b), I have an attitude to this particular action. This means that I approve of it in virtue of some general characteristic it has, and that I should approve of any other action which had this characteristic. But if so the rule 'actions of this kind are right' is not the result of generalisation, but is essential to and implicit in the particular occurrence of approval.

There is another reason for the use of 'attitude' rather than 'feeling'. If I have an attitude of approval towards something it will express itself not only in a class of moral *judgments* but in all sorts of other ways: in decisions, choices, advice, praise, blame, remorse. This also accounts for what would otherwise be an argument against the Emotive theory. It may be said that a historian can pass a moral judgment on a past action without any noticeable feelings, quite dispassionately. But this is because judgments of commendation or condemnation are just as adequate expressions of a moral attitude as are feelings. Anything which can be introspectively recognised as a feeling of approval occurs usually in extreme cases or where other expressions are blocked. For, in a sense, the other expressions are both more natural and the best evidence of this attitude. Actions speak louder than words. A man may never express his feelings about marriage or communism or slavery in moral judgments but you could discover what his attitude is in each case by observing his conduct.

There is one other amendment Stevenson makes to Ayer's type of theory; it is the one he himself regards as of the greatest importance; but, as we shall see, it is doubtful whether it is an improvement on the original theory. The reason for this amendment is a criticism regularly urged against the

Hume-Ayer theory (by G. E. Moore among others). Moore regards it as 'an absolutely fatal objection'.

If, when one man says 'This action is right' and another answers 'No, it is not right' each of them is always merely making an assertion about *his own* feelings, it plainly follows that there is never really any difference of opinion between them; the one of them is never really contradicting what the other is asserting.[1]

It would be like two people saying 'I like sugar', 'No, I don't like sugar', where the word 'No' is obviously absurd. As Moore says elsewhere:

If two persons think they differ in opinion on a moral question (and it certainly seems as if they sometimes *think* so) they are always on this view making a mistake, and a mistake so gross that it seems hardly possible that they should make it: a mistake as gross as that which would be involved in thinking that when you say 'I did not come from Cambridge today' you are denying what I say when I say 'I did'.[2]

Stevenson admits that the argument has great force and devises an addition to the Hume-Ayer theory to meet it. He maintains that a moral judgment has a double function. It describes the attitude of the speaker and it also attempts to impose this attitude on the hearer. This second feature, the 'persuasive' element, is Stevenson's answer to Moore. 'This is right' and 'this is not right' (so far as they both describe the attitudes of the two speakers) are mutally compatible and are both *true*. But what about persuasion? I was asked by a government department to describe the political affiliations and psychological balance of an ex-pupil. I asked two colleagues what they thought about my doing this. One said: 'It would be right' and the other said 'No, it would be quite wrong.'

1. *Ethics*, pp. 100-1
2. *Philosophical Studies*, pp. 333-4

Now on Stevenson's view both of them are trying to persuade *me* and the result is a conflict. The word 'No' becomes quite appropriate. Compare 'Shut the door,' '*No*, leave it open.' Not only is there conflict, but in a sense there is logical contradiction, because it is logically impossible for me to be persuaded by both of them.

Moreover this persuasive element explains why people go on conflicting about moral issues. When it is a matter of taste, we do not mind saying 'tastes differ' and leaving it at that. And indeed the results are sometimes satisfactory; 'Jack Sprat would eat no fat, his wife would eat no lean.' But since moral judgments are intended to be persuasive we cannot agree to differ, any more than rival missionaries can, when they both meet a possible convert. But there is still a fundamental difference between Moore and Stevenson and between this kind of conflict and the normal case of contradiction. When two ordinary statements contradict each other one must be true and the other false. 'Liverpool is bigger than Manchester'; 'No, it is not.' This conflict is resolved by population statistics. But when two 'persuasives' conflict, neither is logically superior to the other. The solution of the Liverpool/Manchester conflict is the prevalence of truth. The solution of a conflict of 'persuasives' is simply victory (and the victory of either side, no matter which).

Nevertheless Moore admitted in a later work that Stevenson's theory went a long way to meet his difficulty and that as a result he was uncertain whether his original view or Stevenson's was correct.[1]

It has also to be remembered that the conflict is solved by the victory of either side no matter which and also *no matter how*. The weakness of duelling as a way of settling problems of honour is that the best (most honourable) man need not win.

---

1. *The Philosophy of G. E. Moore*, pp. 535–54

Truth is merely the ideas which are felt in a certain way and are felt to dominate in a mind or set of minds ... You may indeed ask psychologically, if you please, how they come to dominate, but however they have come to dominate, their truth is the same. If you and I disagree ... and if you argue with me and persuade me that is one way of agreement. But if you prefer to knock me on the head, that, so far as truth goes, is the same thing, except that there is now truth not in two heads but in one. And as to there being any other truth *about* all this state of things, or in short any truth at all, except mere prevalence, the whole notion is ridiculous. And if you deny this you do but confirm it, since your denial (though of course true) must also be false since it is true only because in fact it has prevailed.[1]

Stevenson's insistence on the persuasive element in moral judgments certainly fits some instances of these very well. 'You ought to see it through' is such an example. And perhaps *any* moral judgment thought of as a communication between people can be represented as having this character. So far as moral judgments enter into history or biography they can be regarded as attempts by the author to persuade the reader to adopt the author's attitude to Nero or Nelson or Lloyd George.

But there seem to be equally clearly examples of moral judgment where the persuasive element is lacking. Stevenson admits this, when he says he has 'concentrated on the interpersonal use of moral terms and treated the personal use by implication only'.[2] What is this 'personal use'? It occurs when I make a moral judgment without communicating it to anybody. For example, I may say to myself: 'I ought to see this through', or I may write in my diary (carefully locked away from human eye): 'I treated Jones badly today' or 'Jones treated me badly today.'

1. F. H. Bradley, *Essays on Truth and Reality*, p. 112
2. *Ethics and Language*, p. 134

Now surely these *are* moral judgments, but how can Stevenson argue that they are persuasive as well as expressive? He tries to do this by treating them as *self-persuasions*.[1] I am trying to persuade myself. He asks How do we come to a decision? How do we resolve a moral conflict?' and replies: 'We imagine ourselves to be one of our own heroes; we picture ourselves in conflict with a doughty opponent whom we finally convince. We personify the opposition within ourselves. We call it the devil within us, the old Adam; we exhort it to surrender.'

Now what decision are we trying to reach, what conflict are we trying to resolve, in these cases? Not a conflict or a decision about what is right or wrong. We have already reached that decision; for, if we had not, we could not call the opposition the devil in us or the old Adam. The decision then is a decision what to *do* and the conflict is one between duty and desire (the devil, the old Adam). When I call up images of my heroes or picture myself wrestling with the dragon, I am trying to get myself to do what I have already decided that I ought to do.

But persuasion of *this* kind is irrelevant to Stevenson's analysis. His formula to explain 'x is good' is 'I approve of x; do thou likewise'. Do thou what? Do thou *approve*. And indeed only such an analysis can deal with judgments about third parties or the past. When I say to you 'de Gaulle is behaving badly'. I am trying to get you to *disapprove* of de Gaulle. I am not trying to get you (or de Gaulle) to *do* what you or he approve. When I say to you 'The Treaty of Versailles was unjust' I am getting you to disapprove of it and not trying to get you (or the treaty makers) to do what you or they approve. But when I say to myself or write in my diary 'Jones behaved badly' I am expressing disapproval of Jones and I cannot possibly be persuading myself to disapprove. For why

1. Ibid, pp. 147ff.

should I? Either I *do* disapprove already and persuasion is pointless, or I don't and persuasion is groundless. The cases of Couéism or genuine self-persuasion well described by Stevenson are cases in which I am trying to persuade myself to do what I already approve. There is a story that the undergraduates of an Oxford college during the 1914 War used to share their bathhouse with the Head of their college. And they used to hear him say, in his impressive deep voice: 'Come along now, Phelps. Be a man, Phelps. In you go, Phelps.' This was obviously self-persuasion. But the Provost was clearly not trying to persuade himself to *approve* of taking a cold bath. He already approved and was persuading himself to take it. So I conclude that Stevenson's amendment of the expressive theory to meet Moore's criticism, by the addition of a persuasive element, fails because, in some moral judgments, no such persuasive element occurs.

The truth is surely that moral judgments (like any other judgments) may be used to persuade, as well as to do their proper job. The judgment that the earth goes round the sun is properly and primarily a statement in astronomy about planetary movement. But it could be used as it was (unsuccessfully) by Galileo as an attempt to persuade his persecutors to change their beliefs. And so 'Eppur se muove' takes its place not only in astronomy where it properly belongs but also in social history. So also 'that's nice', said meditatively (and this is the proper use) by the gourmet as he savours his caviare, may also be used by Nurse to persuade Tommy to eat his porridge.

This use of the moral judgment for persuasive purposes is another justification of the substitution of 'attitude' for 'emotion' which was discussed earlier. It means that one of the activities which I shall (on appropriate occasions) pursue, if I have an attitude of disapproval of something, will be to try to persuade others to share this attitude.

There is a final difficulty about Emotive theories, also raised by the previous discussion. Any kind of language may be used for persuasion; and all such persuasions imply approval. How then is *moral* approval to be distinguished from other kinds? This is especially difficult if the theory is one of 'feeling' or 'emotion' and not of 'attitude'. For the only way in which *feelings themselves* can be distinguished from each other is by introspection or direct awareness. It is no use to say that the kinds of things towards which the feelings are directed provide the differentia, for the whole point of an emotive theory is to deny that there is anything specifically moral in the states of affairs which are approved or disapproved. Stevenson seems to be attempting to indicate such a differentiation between emotions when he says[1] that where 'good' means 'morally good' it 'refers not to *any* kind of favour the speaker has, but only the kind that is marked by a special seriousness or urgency'. But this is surely unsatisfactory. Aesthetic approval may be serious or urgent also. Stevenson goes on to adopt in effect the attitude analysis and to appeal to a variety of responses in which the attitude (or disposition) of disapproval may be actualised. The observer may be indignant or shocked and these are clearly different from mere displeasure. The agent may feel guilty or conscience-stricken and these are different from merely feeling annoyed with oneself. And there is a peculiarly heightened sense of security when what is approved prospers, which is not mere pleasure. 'These differences of response ... help to distinguish the attitudes which are moral from those which are not.' But here again the appeal is ultimately to introspection. And it is difficult to see any connection between these different feelings.

It was agreed in the first chapter of this book that the distinction between moral and non-moral uses of the word 'good' is not an easy one. What is here suggested is that the

1. Op cit., p. 90

The Emotive theory

Emotive theory fails to cope with this difficulty satisfactorily. Indeed one of the theories Emotivism was meant to reject was intuitionism—the theory that we just directly apprehend that kindness is good or that cruelty is bad. But Emotivism is itself a kind of intuitionism. We just directly apprehend the difference between feeling moral approval and feeling aesthetic approval or between feeling shocked and feeling displeased.

REFERENCES

A. J. Ayer, *Language, Truth and Logic* (2nd ed.), chapter vi and Preface, pp. 20–2

C. L. Stevenson, articles in *Mind*, January 1937 and July 1938; *Ethics and Language; Facts and Values* (a collection of essays which includes the articles from *Mind* noted above)

9

SUBJECTIVISM AND OBJECTIVISM

Theories of the Emotive type examined in the previous chapter are examples of a subjective view of moral values. Subjectivism in the past has rested on the evidence of variation in moral judgment and the lack of any verification-procedure when moral views differ. This has naturally led to the conclusion that, when views differ, it makes no sense to ask which judgment is nearer the truth, and consequently to the conclusion that moral judgments do not express cognitive states such as knowledge, belief, opinion; for if they did it would make sense to ask about their truth and falsity.

The evidence of variation and non-verifiability has been supplemented in recent philosophy by other considerations making for subjectivism. It has been noted that an objectivist theory has to rest on a number of propositions of a type which modern logic rejects entirely. These propositions assert necessary connections between different characteristics of states of affairs in the world. They are called technically 'synthetic *a priori* propositions'. 'Synthetic' because the states of affairs are different from each other; for obviously one could assert a necessary connection between being a horse and being an animal if animality were part of the *analysis* of the term 'horse'. '*A priori*' because necessary in a strict sense. For scientists may be said to discover necessary connections—between microbes and diseases, between friction and heat. But these connections are not necessary in a strict sense as they rest on empirical evidence (are '*a posteriori*') and have only the degree of probability such evidence justifies.

Some examples of the way in which objectivist theories involve *a priori* synthetic judgments may now be given. Ross argues against the view that moral judgments merely express attitudes of approval. He says that we cannot decide whether or not to approve of something until we have a *reason* to do so; and the reason why we approve of an action is that we recognise in it the attribute of rightness. Now what sort of connection is this, between approving and recognising as right? It is surely a *necessary* connection, and *a priori*. For if it were of the scientific type it would require evidence and it would be possible to apprehend the one characteristic without the other. But what evidence could there be requiring us to connect approving with recognising as right? And it would be odd to say 'I recognise this act as right but I do not approve of it.'

Other examples can be elicited from such a theory as that of G. E. Moore. Moore holds that goodness—intrinsic goodness—is a simple characteristic which certain states of affairs possess. But which states of affairs? Those involving enjoyment of beauty or personal affection. Now the connection between aesthetic enjoyment and intrinsic goodness is, for Moore, a strictly necessary connection, which is just directly apprehended, which is independent of the actual existence of particular people, and which requires and can receive no demonstration. Here then is our first example from Moore of an *a priori* synthetic proposition. 'If any state of affairs includes aesthetic enjoyment that state of affairs is intrinsically good.' The second example which Moore exhibits is the proposition connecting obligation and goodness. 'Ought' and 'intrinsically good' are clearly for Moore different notions, though at one time he failed to notice this; and they are necessarily connected.

If the total amount of good brought into the world by my doing an action would be greater than that produced by any alternative action I could do instead, then it *necessarily follows* that I ought to do this action.

This is Moore's basic principle.

Yet another example of an *a priori* synthetic proposition central to moral philosophy is that already referred to in Chapter 5—'ought implies can'.

In all these cases two different characteristics of states of affairs are said to be necessarily connected. The philosophers who reject *a priori* synthetic propositions divide all propositions into two classes, those resting on empirical evidence and those whose truth rests on our use of language. The only propositions which assert necessary connections are 'analytic'. An example is a dictionary definition. 'Democracy is a form of government' tells us about how the *word* 'democracy' is used. It tells us nothing about the existing world. To discover whether there are any democracies would require observation. Now it is obvious that analytic propositions which do nothing more than *analyse* the meaning of a word do assert necessary connections. A black horse is necessarily black. A triangle is necessarily a plane figure. But there are problems here. A man may know what the word 'triangle' means and say correctly that it is a plane figure bounded by three straight lines. But how many *angles* has it? It is obvious that if a figure has three straight sides it has three angles; but these two characteristics are different yet necessarily connected. If this is thought to be too easy and obvious an example (for, after all, 'triangle' might be said, by its derivation, to include having three angles) another example may be given. What is a cube? It is a solid, all of whose sides are squares. How many sides? Gamblers might well think this also an obvious characteristic *included* in the notion of cube, for dice are a help. But how many edges? To say that *this* is *included* in the nature of 'cube' is unplausible. Hence some of those who reject synthetic *a priori* propositions define an analytic proposition as one asserting the inclusion of one characteristic in another *or the entailment of one characteristic by another*. But it now becomes unplausible to say that analysis deals with *linguistic* facts because entailment of one characteristic by another does not look like a linguistic fact. These are difficult logical problems and cannot be further pursued here. But they lead to a problem which is very relevant to the subject of this chapter.

Those who believe that there are no synthetic *a priori* propositions have to decide what the propositions of philosophy are. And they decide inevitably that they are all analytic. Philosophy tells us no new facts about the world. It tells us how certain words are used; it clarifies their usage and removes confusions and paradoxes into which their use might tempt us to fall. Now most of those who hold this view are subjectivists in moral philosophy; and it will be the object of the following argument to throw doubt on the compatibility of these two views—the 'analytic' view and subjectivism.

Let us return to the Emotivist theory as an example (for Professor Ayer combines the Emotivist view with the rejection of *a priori* synthetic propositions). This view maintains that, when Mr Churchill called Hitler 'that bad man', anyone who really understood the situation would realise that he was not attempting to describe Hitler but was expressing or describing a characteristic of his own mind. And when Dr Goebbels would reply that Hitler was not a bad man he too was saying nothing about Hitler but describing or revealing something about himself. And both descriptions were true. No matter what sort of a man Hitler was, Churchill and Goebbels were fully and equally justified in what they said. For what they meant to assert or express was something wholly independent of any moral facts about Hitler. For there are no moral facts.

Now can such a view of a moral judgment claim to be an *analysis* of it? The analysis of a statement presumably should tell us what we mean when we make it, or what we shall accept as being what we mean when we see the matter more clearly and any confusions are removed.

Now it seems to me that no normal user of an ethical sentence would accept a subjectivist formulation as an *analysis* of it. There is no question of his being muddled or confused. In fact, as I shall try to show, the boot is on the other leg. It is only confusion and muddle that enables a subjectivist theory to pose as an analysis of moral judgment. The more clearly you get anybody to see the meaning and implications of subjectivism, the more emphatically would he reject it as an account of what he meant to say. Imagine yourself trying to persuade Mr Churchill that when he said 'that bad man' he was not describing Hitler and in no way contradicting Dr Goebbels.

Why have subjectivist theories been accepted as analyses of ordinary language? One reason certainly is that language can be used for a great variety of purposes. 'The man's a fool' may be used to express fury with him as well as to state a fact. A judgment concerning spatial location may function as a command. Does this sound wild? 'There is the door' may be one command and 'here is your castor oil' another. The second reason why people have claimed that subjectivism gives an analysis of moral judgments is that they have confused here (as often elsewhere) the meaning of a statement with its truth or with the evidence for it or (as above) with the other purposes for which it may be used or (again above) with the psychological states for whose existence its utterance is evidence.

I may write a letter to *The Times* and say 'Seventeen murderers were hanged in 1947'. In its context this statement evinces a belief of mine; it also evinces an interest in capital punishment. Moreover being written to *The Times* it is intended to influence the attitudes of others. But the meaning of my sentence is wholly independent of these facts about my state of mind when I wrote it. It is true that we sometimes use the word 'mean' so as to include these wider implications. I say 'Jones is a bore'. You reply 'You mean you don't want him asked to lunch'. This is the same use as in 'smoke means fire'. But even here we should probably distinguish between what I mean and what *my words* mean. It would be odd to say that the *statement* 'Jones is a bore' *means* that we don't want him to lunch. Stevenson, it is true, limits the meaning of an ethical judgment to include only those psychological antecedents or consequences of it which would not have been associated with the words used without a process of conditioning.[1] Thus the frowns and praises by which we were originally persuaded to call something 'right' or 'good' are part of the meaning of 'right' or 'good' 'in the psychological sense of meaning which is here in question' provided that the frowns and praises were indispensable.[2] What is the meaning of 'King Charles I was executed'? Suppose I had been so bad at history that I could not be got to learn this fact without lines and beatings; then these indispensable conditioning stimuli would be part of the meaning of the statement. Its meaning would include my sufferings as well as the King's. In any case, the psychological associations of uttering a sentence cannot be discovered by 'clarification' of it and are not entailed by its terms. If this were so, logic and semantics would be engulfed by empirical psychology.

It is doubtful whether Stevenson can be classed as belonging

1. *Ethics and Language*, pp. 57, 61
2. Ibid., p. 69*n*

to the school which holds that the statements of moral philosophy are analytic. He admits that the ordinary man claims objectivity for his moral judgments. But he asks:

> When the confusions of belief and attitude are cleared away and when psychological mechanisms which these confusions have fostered were accordingly readjusted, would people then feel that some more objective criterion is required?[1]

He is saying here that ordinary people do hold an objective view, but that he believes that he could so re-adjust their psychological mechanisms that they would become subjectivists. He is on the way to the view held by J. O. Wisdom that the correct analysis of moral judgments is objectivist but that psycho-analysis of those who assert them could substitute subjectivist judgments for them, as it can cure people of a belief that they are being persecuted.

A man says that there are fairies at the bottom of his garden. You are entitled to say to him: 'Your only evidence—that flickering blue light—is compatible with an escape of marsh-gas for which there is other evidence about here.' He might be shaken or even convinced. But you cannot now say to him 'When you said there were fairies down your garden, you *meant* there was marsh-gas down your garden'. Or, moving to an argument which has been held to be philosophical, we may take the example of colours. The plain man certainly clothes objects in their qualities. Like Berkeley he is convinced that snow is white and fire hot. It is not too difficult to convince him that the colours he sees are not 'in' or 'on' the objects, but are 'in' his retina or his pineal gland or his sense-data or his mind. But it will not do to say that when he said 'snow is white' he *meant* that light is falling on its surface and all the radiations are being reflected from it to the retina where ... etc., etc. Or, to take Wisdom's case, a man may be cured of a

---

1. Ibid., p. 31

persecution complex by psycho-analysis and come to agree that he is not being persecuted. But it will not do to say that when he said (before treatment) that he was being persecuted he *meant* that he was not being persecuted. Similarly with morals. It is very easy (all too easy) to turn a beginner from being an objectivist to being a subjectivist (by such arguments as those from the variation of moral judgments and the impossibility of verifying them). But even the most stupid beginner is usually quite clear that he was not a subjectivist when he sat down in your armchair for treatment.

I conclude then that subjectivist theories of morals are not compatible with the analytic theories of philosophy with which they are so often associated. The most a subjectivist can say is that an objectivist attitude is 'built in' to our use of moral language. We believe in the objectivity of 'good' and 'right', as ordinary men believe in the objectivity of 'white' and 'hot'. But these beliefs are simply false. There is a standing tendency to error in the ordinary use of moral language as there is in the ordinary use of colour language. This error can be corrected by the traditional arguments for subjectivism urged long before the linguistic theory of philosophy was advanced, and cannot be corrected by linguistic considerations, all of which point in the opposite direction.

A further difficulty for subjectivism is this. Many subjectivists fail to go the whole way with their theory, and thus give reluctant and unintentional support to the objectivist tendencies which are so hard to eradicate. A preliminary example may be taken from the closely parallel field of aesthetic language. Here, too, ordinary people hold strongly to objectivist usage. They may think there are doubts about whether the paintings of Picasso and the music of Schoenberg are beautiful; but they regard this as a genuine question and not one to be removed entirely by saying that it is all a matter of individual feeling.

It is true of course that some men have much experience of

the variations of aesthetic judgment, and experience for themselves the impossibility of *proving* to anyone that a picture they love is really beautiful. And some of them do in fact become subjectivists as a result. They use aesthetic language quite differently. They do not contradict people who differ from them. Instead of saying 'That picture is a daub' they say 'That is not the kind of picture I fancy'; or they say 'Pope may please other people but he says nothing to me' or 'Schoenberg is not *my* kind of music'. But the subjectivist must be consistent. My example of a failure here is from Anatole France.

There can no more be objective criticism than there can be objective beauty; and anyone who thinks that he puts into his appreciation of art anything but himself is a dupe of the most fallacious sophistry. We must recognise that we speak only of ourselves when we have not the self-control to remain silent. The good critic is one who recounts the adventures of his own soul as it voyages among masterpieces.

In the last sentence the key words are 'good' and 'masterpieces'. To indicate the difficulty, I may quote a speech delivered in the legislature of Georgia by Mr Hal Wimberley.

There are only three books in the world worth reading: the Bible, the Hymnbook and the Almanac. Read the Bible; it teaches you what to do. Read the Hymnbook; it contains the finest poetry ever written. Read the Almanac; it teaches you to figure out what the weather is going to be. There is not another book which it is worth anyone's while to read and therefore I am opposed to all libraries.

Or again one of the gems in my collection of bad poetry is a slim blue volume which I forbear to name. The volume carried with it a flysheet of review notices. 'Contains good poetry and some very nice thoughts' (*Hertfordshire Mercury*);

'Excellent verses arranged with touching delicacy' (*Galloway News and Kircudbrightshire Advertiser*). Now, on the subjectivist view, it makes no sense to say my slim blue poet is a *bad* poet. Nor can the adventures of the soul of Anatole France claim any precedence over the still more astonishing adventures of the souls of Hal Wimberley and the *Hertfordshire Mercury*.

So also in ethics. Philosophers who deny all absolute values and say values are made in individual choices are found asserting absolute values themselves. The existentialists reject all rules and abstract propositions about duty or goodness. Each man makes his values as he makes his choices. Yet we find constant emphasis on the essential value of 'commitment' (Kierkegaard) or 'engagement' (Sartre). And the existentialists echo a very common 'subjectivist' view which one often hears among ordinary men. 'We differ on every issue but I respect his *integrity*.' (So we do *not* differ on *every* issue.) This universal and absolute value is common ground to the existentialists. 'Authenticity' (Kierkegaard and Heidegger), 'fidelity' (Marcel), 'sincerity' (Sartre) are other names for it. And the effect of many individual standards is 'tolerance', but that is itself not an individual but an absolute standard.

So too with British philosophers. One of the earliest upholders of an 'approval' Emotive theory of morals was Adam Smith. For example:

> To approve or disapprove of the opinions of others is acknowledged by anybody to mean no more than to observe their agreement or disagreement with our own.[1]

Yet he recognises the greater insight of some judges in morals as in aesthetics.

> It is the acute and delicate discernment of the man of taste, who distinguishes the minute, and scarce perceptible, differences of

1. *Theory of the Moral Sentiments* (1759), p. 24

beauty and deformity; it is the comprehensive accuracy of the experienced mathematician, who unravels, with ease, the most intricate and perplexed propositions; it is the great leader in science and taste, the man who directs and conducts our own sentiments, the extent and superior justness of whose talents astonish us with wonder and surprise, who excites our admiration and seems to deserve our applause.[1]

This is not subjectivist language nor compatible with the sentence previously quoted concerning approval.

There is a similar inconsistency in a recent presentation of the subjectivist theory by Professor W. H. F. Barnes.[2] Barnes says that I answer moral questions by consulting my own moral feelings. But suppose these change? It does not make sense to say they have *improved*, for that would suggest a standard beyond them. If Barnes says his moral sensibility has improved this simply means that he himself approves more of its later than of its earlier judgments. And if I say he has progressed I mean that I approve more of him now than then. (Compare the first quotation from Adam Smith above.) But Barnes talks of *refining* his moral attitudes[3] and of '*a moral sensibility awakened to new facts*' and of '*undeveloped* moral sensibility'[4]. I find it difficult to see how these phrases can be compatible with complete ethical subjectivism.

So too Miss Margaret Macdonald, writing about 'Natural Rights', says that 'there are no true or false beliefs about values but only better or worse decisions and choices'.[5] Yet she also says 'a criminal cannot be cast out of humanity',[6] and 'It is in

1. Ibid, pp. 32–3
2. *Proceedings of the Aristotelian Society*, suppl. vol. XXII
3. Ibid., p. 24
4. Ibid., p. 27
5. *Proceedings of the Aristotelian Society* (1946–7), p. 250 (reprinted in Laslett, *Philosophy, Politics and Society*, p. 54)
6. Ibid., p. 246 (p. 51)

some sense true that no one ought to be ill-treated because he is a Jew or a Negro or not able to count above ten.'[1] Yet on her own view there is nothing true or false here. These sentences record only her own decisions. Hitler and the Ku Klux Klan decided contrariwise. Therefore it is equally (in some sense) true that Jews and Negroes should be victimised. And how can there be better and worse decisions? Is the statement 'Miss Macdonald's decision is better than Hitler's' a moral statement? If so, it simply records the decision of a third party or Miss Macdonald's reiteration of her own.

The subjectivist view also makes a great deal of our ordinary talk pointless or misleading. If I say of a man that his moral judgments are bigoted, perverted, or prejudiced, I can mean only that they differ from mine. When I say 'Jones has reformed during the year' I can only mean that I disapprove of him as he was last year and approve of him as he is now. I cannot mean that his moral standards have changed for the better. There can have been no moral change *in him*. Of course it may be said that, when we say of a man that he has reformed, we do not mean that his moral standards have changed, but that he has begun to live up to them, as he did not do last year. But this leads to the even more unpalatable conclusion that this change too is not a moral change *in Jones*. It is just that I approve of people who live up to their own moral standards. To say this is a moral merit *in them* is mistaken.

Moreover, if there is no truth in these matters any difference between me and one man in our moral estimates should matter as much or as little as any difference between me and any other man. But what I find is that some such differences shake me and some leave me unperturbed.

A disagrees with me. A is a man I respect. He lives up to his own highest standards. I hesitate to claim that I have as much right to my view as he to his. B disagrees with me. He

[1]. Ibid., p. 243 (p. 49)

is a man for whom I have no respect; he makes no effort to live up to his own standards. I don't mind his difference from me. C differs from me. C is a man of moral scruples. He has often noticed some subtle moral issue which I might have missed. He sees the relevance of related problems. My difference from him worries me. D differs from me. He has regularly shown himself a man of no discrimination at all. He sees no difference between different cases and no relation between similar ones. He would not steal the railway company's cutlery, but he thinks it quite all right to travel without a ticket. He defends free enterprise as a recognition of the value of the individual and treats his staff in a way that it would be polite to call feudal. My difference from him causes me no worry. E differs from me. E is a man of wide experience of men and affairs. He is of the world without being worldly. I hesitate to differ from him. F differs from me. He is a narrow man, shut off by upbringing and vocation from human contacts, with a nature obviously warped and embittered by these limitations. I have no hesitation about differing from him. G differs from me. He is a man of good practical judgment. His opinions about people and policies, whenever they can be checked, are highly reliable. He is alert, critical, and sensible in human relations. I wonder whether he is not perhaps to be relied on in moral matters too. H differs from me. His judgment on practical affairs is normally bad. He is silly and impractical in choice, credulous and superstitious in belief, hasty and rash in his judgments and predictions about people and policies. I naturally expect him to be an unreliable moral judge too.

What am I doing in making these contrasts? Just what I would do if I were on a jury in a case where there is no direct evidence and two witnesses disagree. I ask myself which witness seems the more trustworthy. Of course this is a risky procedure. Sometimes the most candid and best qualified witness is wrong and the most shifty and ignorant witness right. Nevertheless, in the absence of other decisive evidence, my procedure is rational. And it presupposes objectivity. For where there is no truth no one can be more trustworthy in judgment than anyone else. These contrasts are insignificant for, if 'good' and 'right' signify nothing but the approval of the speaker, there is nothing to choose between A and B, C and D, etc.

The parallel with aesthetic judgment may help us again here. Variations in judgments of beauty are even more striking and obvious than in judgments of right and wrong. They have led to subjectivist theories of beauty. We have seen in the examples of Anatole France how difficult it is to keep out all objectivist elements. Here too, as Anatole France suggests in his talk of 'the good critic', the same issue arises. While variations do occur we take more seriously the judgment of a man who has spent his life with pictures, whose 'attributions' are later verified, who can spot late additions, than those of someone who merely 'knows what he likes'.

In art we are equally tempted to use the language of Professor Barnes and talk of 'improving' or 'refining' one's 'sensibility'. In art too I do not see how one can talk like this without implying that the more refined sensibility is capable of better judgment. It might be said that one man may be more sensitive to pain than another but that this does not mean that there is one objective pain, which they are both assessing. It only means that one man feels pain more acutely than another (perhaps from a similar stimulus). But we do not mean this when we speak of aesthetic or moral sensibility. The difference is not here a quantitative one (though it may also be true that a good critic's enjoyment or distaste is in fact greater than that of the ordinary man).

The alternative to subjectivism both in aesthetics and in morals is to claim that men are endowed in varying degrees with a capacity for aesthetic and moral appreciation and discrimination. It is a mistake perhaps to call this faculty a *rational* faculty because of the difficulties of verification and proof which we examined earlier. It is perhaps equally mistaken to react to the opposite extreme and call it a '*sense*'. We have a word for this kind of faculty in other fields and it is the word 'judgment'. Judgment is exercised in a wide variety of ways. A manager may be appointing someone to a position of responsibility; a selection committee may be interviewing candidates for admission to a college, for a service commission, or for the treasurership of a charity; a juryman may be estimating the reliability of two conflicting witnesses; a farmer may be deciding whether to enlarge his holding; an investor whether to move into equities at once. It may be said that in all these cases there may be later evidence that the judgment in question was sound. The departmental head goes on to further successes, the selected candidate gets his first class or a staff college job in due course, the farm prospers, the investments double their yield. But in morals and aesthetics there is never a pay-off like this. While this must be granted, there are still sufficient similarities between the actual processes of judgment in the verifiable and the unverifiable cases to make the differences seem irrelevant or at least much less relevant. In all cases an alert mind, an attention to relevant detail, the holding together of a great variety of different 'indications', all these result in an 'impression' of reliability, potentiality, integrity, propriety, 'rightness' in each field. Everybody knows how fallible these processes are, how they may be dismissed as 'hunches' or 'intuitions', and how resistant they are to any formulable type of inference. Yet everyone knows even better, who has worked on such matters, that people do differ enormously in this capacity of judgment; and that the 'hunch' based on a single meeting a year ago or the 'impression' left by 'the look in his eye' is *different* from the 'judgment' resulting from long experience, from working alongside a man (or fighting alongside him), though there is in neither case anything that could count as 'strong evidence' or a case which would convince an impartial spectator.

So it seems to me that a Board of Directors may have two items on its agenda: (a) whether to retire its general manager or keep him on for two more years, (b) whether to dismiss its treasurer who has been detected in some questionable activities with its funds. I should expect these two decisions, the first an economic and the second a moral one, to be reached by discussions remarkably similar in their general pattern—the citation of relevant considerations (with some debate on whether they are relevant) the weighing against each other of pros and cons which can have no quantitative or measurable valuation, the attempt by each director to pull together these variegated threads into a decision, the tendency to be influenced, in the first decision by colleagues with long experience, in the second by colleagues of notable integrity; and a final verdict which will be felt to be justified at least partly by the care and attention spent on it.

So I conclude that, powerful and popular as the subjectivist theory now is, there are insuperable objections to it as an 'analysis' of ordinary language, and some considerations against it and in favour of an objectivist account of the values this language attempts to express.

REFERENCES

(a) Synthetic *a priori* propositions in morals:
   (i) Right and approved: W. D. Ross, *Foundations of Ethics*, p. 23
   (ii) Good and aesthetic experience: G. E. Moore, *Principia Ethica*, pp. 188–9
   (iii) Ought and good: G. E. Moore in Schilpp (ed.), *The Philosophy of G. E. Moore*, pp. 554–71

(b) The logic of *a priori* propositions:
Analytic but including entailment, which is linguistic: A. J. Ayer, *Language, Truth and Logic* (2nd ed.), pp. 16–18 (with references)
The cube and its edges: Langford in *The Philosophy of G. E. Moore*, p. 327; accepted by Moore, pp. 575–6, 599

(c) Subjectivism not an analysis of moral judgments:
J. O. Wisdom, *Proceedings of the Aristotelian Society* (1935–6)
C. D. Broad, *Proceedings of the Aristotelian Society* (1944–5)

10

FREEDOM OF CHOICE[1]

In an earlier argument we used the principle that 'ought implies can'. If a man is forced to do something he can neither be praised nor be blamed for doing it. 'I could not help it' is always a complete defence. But this raises the very difficult problem of free will. The problem has been carefully formulated by Henry Sidgwick. 'Is my voluntary action at any moment completely determined by (i) my character as it has been partly inherited partly formed by my own past actions and feelings, and (ii) my circumstances, or the external influences acting on me at the moment? Or not? Could the volition I am just about to originate be certainly calculated by anyone who knew my character at this moment and the forces acting upon me? Or is there a strictly incalculable element in it?'[2]

During the two past centuries, up to 1900, there had been a sharp opposition between science and free will. Science maintained a rigid determinism and a belief in universal

---

1. This chapter and the next incorporate material from an article on 'Freewill' for the *Encyclopaedia Britannica* and a contribution, on *Freewill and Punishment*, to *Contemporary British Philosophy (third series)*, with the kind permission of the editors and publishers of these volumes
2. *Methods of Ethics*, p. 46

causation, which rejected free will as it rejected miracles. It is true that no scientist would defend the crude fatalistic view, that my action or my 'fate' is determined by forces wholly independent of me, so that it is bound to come to me no matter what I do to prevent it. This is the attitude of the soldier who says 'The bullet which is going to kill me has got my number on it, so it does not matter whether I take cover or stand up in full sight of the enemy'. This type of fatalism, sometimes called 'oriental', is wholly unscientific: for every scientist would insist that the behaviour of the object in which a change occurs must always be one of the factors contributing to the causation of the events which happen to it. So what the soldier does must make a difference.

But the prevalent scientific view was that all observable events are subject to scientific law and therefore completely determined and in principle completely predictable. It was inconceivable that the behaviour of a single species on a minor planet in one among countless solar systems should escape a type of determination which had been succesfully found to apply on the widest scale of stellar magnitude and to the smallest microscopic bodies. It was true that up to 1850 the successes of science had stopped short at inanimate matter. But from that time onward Darwin in biology, Marx in sociology, Pavlov and Freud in psychology were advancing causal explanation across the frontiers of life and mind.

In the last fifty years, however, the scientific atmosphere has greatly changed. 'Determinism' is no longer an acceptable scientific term. There is little talk in science of 'irresistible forces' or indeed of any 'forces' at all. To talk of magnetic *attraction* is unscientific for the force of attraction is completely unobservable. What can be verified, and therefore should be said, is that, in the vicinity of certain bodies, iron filings move in certain paths or arrange themselves in certain patterns. There is no 'compulsion' here. What is observed is regular conjunction and no more. And again in this regularity there is no observable *necessity*? Scientists no longer *explain why* events occur. They *describe how* they occur. Nevertheless, though the assertions of compulsion and necessity disappear, the claim to predict still remains. Yet it is also noteworthy that the predictions are not endowed with certainty but only with that degree of probability that the evidence justifies. Yet even this would still seem to threaten free will. For prediction with a very high degree of probability that a man will choose A seems to threaten his freedom to choose A or B.

There are, however, some special features of human experience which make even these milder claims of modern science of doubtful application to it. There are three features of scientific method which seem essential for any close approximation to accurate prediction. The first is measurement. There are always differences between the specimen whose behaviour is to be predicted and the specimens whose behaviour is evidence for the prediction: differences of mass, of chemical constitution, etc. These differences are accommodated by having laws with variables, to which quantitative values can be given from the measured features of the particular case brought under the law. In order to predict from the law of gravity how body X will behave in relation to body Y the distance between these bodies must be inserted as a determinate element in our calculations. Now the psychological phenomena from which human conduct would have to be predicted are not measurable though they may be quantitative in intensity. It makes sense to say that I am feeling more angry or suffering more pain than a minute ago. But it does not make sense to ask whether my pain has doubled or my anger gone up by nine per cent. The only way to try to measure psychological phenomena is to measure their physiological accompaniments. (For pain, twice as many tears or twice as many decibels in the groaning!) But to predict conduct wholly from physiological data is to deny either the existence or the causal relevance of psychological states. The former denial (that involved in materialism or behaviourism) is now generally discredited. The latter, holding that psychological events are 'epiphenomena', accompanying but making no difference to physiological processes, is difficult to reconcile with evolutionary theory. For such 'epiphenomenal' states would be useless to the organisms which manifest them and would have been expected to atrophy rather than to increase in number and complexity as they have done.

If, then, psychological states are not quantitatively measurable, prediction is impossible. For example a psychologist may establish that a child who has been shut in a dark cupboard will subsequently tend to fear being shut in. But he cannot give any quantitative value to the fear and consequently cannot predict whether in any particular case a rival factor such as hunger will overcome it.

The other two features of scientific method which make it inapplicable in full rigour to human conduct are analysis and repetition. By 'analysis' is meant the power to deal with one feature of an object at a time and to regard some features as wholly irrelevant to the calculations required. For calculations of motion, mass and shape are relevant, chemical constitution is irrelevant. For spectroscopic behaviour, chemical constitution is relevant, mass and shape irrelevant. But, if no features of a substance could be thus regarded as irrelevant to the establishment of a law concerning its behaviour, then no such laws could be established. There are grounds for holding that such abstraction is impossible (or possible only in a very rough and ready fashion) in regard to human beings and their psychological characteristics. The features of a man's character are so interwoven that none of them would be what it is without the others and these features correspond to the forces or features of a scientific object which go to determine its

Freedom of choice

behaviour. We tend to predict the actions of men by establishing that they have certain dispositions such as courage, wit, intelligence, etc. But a man's quick temper helps to determine his kind of courage; his wit gives his temper its peculiar edge; his intelligence lightens his wit; his sensitivity broadens his intelligence; his imagination extends his sensitivity, and so on. Even this language is misleading for it suggests that these features exist separately and *then* affect each other. But the truth is that to understand any one of them is to find it stamped with the man's individuality. The unity of the self makes any attempt to abstract or isolate psychological components impossible. Such dissection spells death. So, in writing testimonials or references, I find that, when I do not know a man well, such general words as 'impetuous', 'tolerant', 'vigorous' seem to describe him very well. But the better I know him the more utterly they fail to convey the person I know. Kierkegaard makes the same point: 'It is more difficult to describe one actor than to write a whole philosophy of art . . . The more one can depend on generalisations the easier it is, for the material is so vast that all the completely abstract observations, which anyone can learn by heart, seem to mean something. But the more concrete the observation the more difficult it is.'[1]

The third feature required for scientific prediction is repetition. By this is meant the assumption that any feature isolated within a single specimen can be expected to recur in other specimens with a difference which is either irrelevant or expressible in quantitative terms as values of the variables in the scientific law which governs the inference. But each human personality is unique. Study of pieces of sodium provides good evidence for the behaviour of the next piece of sodium. Study of the behaviour of Smith, Jones, and Robinson is poor evidence for the behaviour of Brown. This is recognised by those who have devoted themselves to any field of human

1. *Journals 1834–1854*, English trans., p. 147

experience. A few quotations out of a great many possible sources will illustrate this. They are all protesting against the attempt to bind human action by rules and formulae.

Martha Graham, a choreographer of genius, said 'There is a vitality, a force, an energy, a quickening that is translated through you into action; and because there is only one of you through all of time this expression is unique. If you block it, it will never exist through any other medium and will be lost'.[1] Benedetto Croce, the great critic, writing on the attempt to establish rules for the creation of beauty, refers to 'erroneous modes of criticism' which ask whether a work of art 'obeys the *laws* of epic tragedy, of historical painting or landscape . . . Artists have always disregarded these laws. Every true work of art has violated some established law and upset the ideas of the critics, who have thus been obliged to broaden the laws until finally even the broadened law has been proved too narrow, owing to the appearance of new works of art, naturally followed by new scandals, new upsettings—and new broadenings'.[2] T. E. Lawrence had studied the textbooks of strategy and the campaigns of Alexander, Caesar, and Napoleon. Lying out in the desert one night thinking about his campaign against the Turks he meditated on what he had learned.

> In military theory I was tolerably read, my Oxford curiosity having taken me past Napoleon to Clausewitz and his school, to Caemmerer and Moltke and the recent Frenchmen . . . my interest had been abstract, concerned with the theory and philosophy of warfare.
>
> Now, in the field, everything had become concrete, particularly the tiresome problem of Medina; and to distract myself I began to recall suitable maxims on the conduct of modern scientific war. But they would not fit and it worried me.
>
> The algebraical element looked to me a pure science subject to

1. Speaking to Agnes de Mille. Cf. *Dance to the Piper*, p. 307
2. *Aesthetic*, English trans., p. 37

mathematical law, inhuman. It dealt with known variables, fixed conditions, space and time, inorganic things like hills and climates and railways, with mankind in type-masses too great for individual variety ... It was essentially formulable ... But a line of variability, Man, persisted like a leaven through its estimates making them irregular. . . . Nine tenths of tactics were certain enough to be teachable in schools, but the irrational tenth was like the kingfisher flashing across the pool and in it lay the test of generals. It could be ensued only by instinct (sharpened by thought practising the stroke) until at the crisis it came naturally, a reflex.[1]

Finally A. E. Housman the leading Latinist of our day writing about textual criticism, that is the amendment of the texts handed down to us, in cases where it seems they are 'corrupt' and their readings mistaken.

Textual criticism is not a brand of mathematics nor indeed an exact science at all. It deals with a matter not rigid and constant like lines and numbers, but fluid and variable; namely the frailties and aberrations of the human mind, and of its insubordinate servants the human fingers. It is therefore not susceptible of hard and fast rules. It would be much easier if it were; and that is why people try to pretend that it is, or at least behave as if they thought so. Of course you can have hard and fast rules if you like, but then you will have false rules and they will lead you wrong, because their simplicity renders them inapplicable to problems which are not simple but complicated by the play of personality.[2]

There is one special difficulty in regarding human conduct as completely determined and completely predictable. Human thought is part of human conduct and the moral or scientific or philosophical beliefs of any man would have to be regarded as predictable consequences of his causal situation. While this kind of explanation is often welcomed by any thinker as

1. *The Seven Pillars of Wisdom*, 1940, p. 193
2. *Proceedings of the Classical Association* (1916), p. 68 (reprinted in A. E. Housman, *Selected Prose*, p. 132)

accounting for the errors of his opponents, he would not like to accept it as applicable to the truth of his own view.

A Marxist will maintain that there are no objective standards of moral or political belief. All human conduct is determined by economic processes and ultimately by the class structure of society. A man may think he is making a choice and deciding on a course of action because it is right. This is an illusion. All his decisions are determined, and determined by class interest; what he calls his moral or political beliefs are those rules which enable the class to which he belongs to acquire or maintain domination in his society. For Marx, the Benthamite *laissez-faire* theory of economics has no rational validity. It simply reflects the interests of the bourgeois industrialists who dominate capitalist society. But if so Marxist economic determinism has no validity either. It also is an illusion on its own showing. It is the inevitable by-product of material forces, of local and temporary class interests, bound to be dialectically destroyed as history develops. Nineteenth-century working men necessarily had to believe it but there is no reason why anyone else should.

It may be objected that any attempt to shut off an area as impervious to scientific treatment is a counsel of despair. Failure has always attended such exclusions; science has swept in and conquered. The area of human conduct is, moreover, an insignificant one on a world scale. Why in this tiny field should laws fail and prediction falter? Perhaps the answer is that only in this tiny area (so far as we know) can such questions be asked. Knowledge of the universe is already so astonishing an exception to the general pattern that it would be even more surprising if other special features did not distinguish men's lives. It is a surrender to sheer size that makes us think the spiral nebulae a more impressive wonder than the minds of Newton and Einstein.

The main arguments here advanced, however, do not shut

off any area from scientific enquiry. Psychology, sociology, and economics all rightly cover human conduct. All that is said is that they will have a lower probability in their predictions than other studies and that their attempts to deal with individual cases will be liable to considerable margins of error. They will succeed so far as quantitative physiological behaviour provides clues to psychological processes, so far as elements in the self can be safely isolated from the rest, so far as men resemble each other in the characteristics relevant to the enquiry and differences do not affect the result. Shylock was right in his evidence, 'If you prick us, do we not bleed? if you tickle us, do we not laugh? if you poison us, do we not die?' But dubious in his conclusion, 'if you wrong us, shall we not revenge?' There is good reason to believe that in the parts of conduct which count most (parts other than pricking and tickling, blinks and knee-jerks) these conditions are largely unfulfilled. Nevertheless, for large-scale statistical purposes such as life insurance or advertising, scientific prediction may yield a result both informative and useful.

If we ask for positive evidence for free will some would answer that every man has direct experience of it. Just as he knows he is in pain, he knows as directly and certainly that he is making a free choice between, for example, going to the cinema and digging in his garden. This claim as it stands must be rejected, for it is difficult to see how anyone could be directly and immediately aware of the truth of the proposition that his decision has no determining cause. But it may certainly be said that most men *believe* this proposition and that this belief is ineradicable. It may be said that it does not follow that the belief is true and that its prevalence, even if it were false, might be explained by its biological utility. As Ross has said, a man is more likely to work for an examination and therefore to pass it if he believes it is uncertain whether he will pass or fail. If he thought either result was certain he might well slacken his efforts (as voters for safe seats or the defenders of forlorn hopes are apt to do). Yet we also saddle ourselves with burdens of guilt or remorse which are linked with our belief in freedom and responsibility, and these burdens would seem to hinder effective action, and to be useless illusions and therefore biologically puzzling, if they are in fact false.

The strongest evidence for free will is indeed the link with moral judgment. 'Ought implies can.' If I cannot help doing X, I cannot be blamed or praised for doing it. If I cannot do X, X cannot be my duty. Now some philosophers have attempted to maintain that this linkage between moral responsibility and free will is not defensible, and even that such responsibility requires a determinist view of human action. Suppose it were the case that, everything being the same including my own self, two alternative actions were possible (e.g. a brave or a cowardly one), then neither could be *my* action as issuing from my character. Choice would be equivalent to chance. If I were to give a reason why I acted in a certain way (that I love Bach's music, or that I was hungry, or that I felt I ought to help the victim of persecution) this could not be *the* reason: for, given that exact degree of admiration, hunger, or pity, I could still have acted otherwise. Mental life for the free will believer is a series of unconnected, inexplicable, unmotived choices.

On the other hand, as is argued, for example, by F. H. Bradley, there is a kind of determinism which is morally acceptable. That is determinism of action by the self as a whole. If anyone predicts my action on the basis of statistical enquiry, I am properly indignant. If anyone says 'I know what you will do because I have calculated exactly the strength of the rival forces battling within you, your desire for praise, your ambition, your addiction to gambling; and the answer comes out as follows', I laugh at this claim. But suppose he says 'I know you well enough to know you could not have done that' or 'You are one of our most reliable members; I knew I could count on you', I accept and welcome this kind of prediction because it is based not on statistics nor on abstractions but on knowledge of *me* as a person.

It does seem true that this kind of prediction based on experience of a person as a person is the only kind that comes within sight of recognising human individuality and reconciling this recognition with the claims of moral responsibility. Yet it too has serious difficulties. It claims that prediction of action based on knowledge of character is both legitimate and acceptable to any self-respecting person. But it is *acceptable* only if the action predicted is a good one. Suppose someone says 'I knew you well enough to be sure you would not say anything; you are too afraid of unpopularity' or 'I was sure you would keep us all waiting'. In such cases the very word 'reliable' would seem like irony or a joke. In welcoming those other predictions I was confusing prediction with praise. It was the latter I really welcomed.

Moreover prediction from knowledge of a person has its own pitfalls as a rational process. It assumes relevantly similar situations and, more seriously, it implies that character cannot be changed by actions and that people cannot learn by experience, especially experience of their own mistakes. There can be no adequate evidence of what this *changing* self will do. Winston Churchill has described how the German ambassador to London was recalled to Berlin for not having foreseen what Lloyd George would say in his important Mansion House speech in 1911. Mr Churchill comments: 'How could he know what Mr Lloyd George was going to do? Within a few hours before his own colleagues did not know. Working with him in close association, I did not know. Until his mind was definitely made up he did not know himself.'[1]

Another argument for relating determinism to moral res-

1. *The World Crisis*, pp. 49–50

ponsibility is that of R. E. Hobart. The free will theory maintains that freedom of choice is necessary if we are to praise or blame anyone for his actions. On the contrary, argues Hobart, determinism is necessary. For we praise or blame a man for his actions; we praise the self whose act it is. We say 'that was a brave deed' or equally 'that was the deed of a brave man'. But here there is a link between the act and the man. If it were possible, given the brave self, for a brave or a cowardly act to be done, then the act cannot be attributed to the brave self but only to the free will. Hence there must be a necessary causal connection between the self and its acts, for praise or blame to be possible. The difficulty with this argument is that the connection between courage and the brave act, while necessary, is not causal. Courage is a *disposition* or tendency. Causes precede effects but the courage we attribute to the man is not a courage he had just before he acted. The courage was *manifested in* the act. We say that rubber stretches because it is elastic. But it is not because the rubber *was* elastic that it *now* stretches. To say rubber is elastic just means that if you pull it it will stretch; if you stamp on it it will flatten; if you twist it, it will be distorted. So of course there is a necessary connection between stretching and elasticity but not a causal one.

Another argument for determinism was originally put forward by Schlick and has recently been fully worked out by P. H. Nowell-Smith and F. Ebersole. It too works on the link between praise and blame on the one hand and moral responsibility on the other, the link which was previously supposed to require free will. The problem is to find a criterion for 'free' action. The answer on this theory is that those actions are 'free' which can be influenced by praise and blame (or by punishment). A schoolboy is not reproached or punished for stupidity because blame and punishment cannot make him less stupid. He is blamed or punished for laziness because a talking to or punishment may make him less idle.

To say that idleness is a moral failing and idle actions are actions a boy 'could help doing' is just to say that idleness is amenable to blame and punishment. 'If it is true that praise and blame are means employed to bring about good events and prevent bad ones, they are appropriate not to all good and bad events but only to those they can in fact bring about or prevent. Since a moral action is one that can fittingly be praised or blamed, it follows that a moral action is one which can be brought about or prevented by those means.' Thus a moral action is not an action without a determining cause but one which is or can be affected by praise and blame and punishment.

The first difficulty about this view concerns the effect of blame or punishment which it requires. As the quotation shows, it must *prevent* wrong action. Ebersole sticks to this strong line. 'We must be able to know that the condemnation will reform. It must fulfil its intention to prevent repetition of the wrong for which the person is punished.' But he has to admit the difficulty here: 'In the strong sense of "justification" which I have been using punishment is never justified. We are forced to take action on the best knowledge available; attempted reform more often than not does not come off.' But surely a justification which can never justify any actual punishment cannot be worth discussing. Nowell-Smith, however, sometimes moves to a less extreme view. 'Fear of punishment will *affect* the future behaviour of the thief . . . there will be a motive *tending to* make him refrain.' 'The point of blame is to strengthen some motives and weaken others.' Here then we shall hold that a man is responsible for an action if punishment would make it less likely that he will do it. But the criterion is now too mild to delimit responsibility. A kleptomaniac does not steal when a policeman or shop assistant is watching him and he takes ingenious steps to avoid detection. His behaviour is therefore affected by fear of punishment. The reason for regarding him as irresponsible is not that this fear fails to affect him but that he shows other signs of motiveless irrationality. He steals one kind of article only—silk stockings or suspenders—and then proceeds to hoard them and not to resell.

The second main difficulty about this view is that the act which is affected by blame or punishment is never the act for which the blame or punishment is awarded, since the one always follows and the other necessarily precedes the award. This is not always clearly recognised. For example in the passage quoted above: 'Since a moral action is one that can fittingly be praised or blamed it follows that a moral action is one that can be brought about or prevented by these means.' But when the praise is given the action it is to bring about cannot yet have occurred. So this action cannot be the action which is being praised. Ebersole again takes the logical way out. The only relevance of the act for which (as we inaccurately say) man is punished is *evidential*. It shows us the sort of man we have to deal with and we can deal with him accordingly. 'Our justifiable concern is with the person as he has been made by the result of his wrong choice. Our concern with his former self is justifiable only because, without enquiring what he was like before, we do not have any way of knowing what he is like after the condemned action.' But surely previous commission of the wrong is not the only evidence we may have. We may sometimes know that a man was baulked in his attempt by external circumstances. The rope broke, the gun misfired. And in such a case Ebersole would have to say that the punishment was equally justified. His own extreme imaginary example confirms this admission. 'If we possessed a sensitive brain-wave machine which would predict whether a person would commit a crime and what sort of censure or punishment would prevent it, then we should have no need to be concerned with a person's past.' We shall be dealing with

Freedom of choice

punishment in more detail in the next chapter; but it is surely obvious that it is impossible to blame or punish someone who has done no wrong.

We noted previously (p. 99) that the view of philosophy as linguistic analysis cannot be combined with subjectivism. Here again there is a conflict. Ebersole agrees that the account he gives of 'ought' and 'moral' and 'free' involves a complete break with ordinary usage. For usage is indeterminist and rejects blame or punishment when no wrong has been done. But a change of usage to determinism and a reform theory of blame and punishment would not be merely linguistic. It would not indicate a triumph of philosophers in persuading man to use language in a way more fitted to the unchanging facts. If language changed, it would be because the facts had changed; since the facts in question are men's actions and attitudes. If deterministic and reform language were adopted it would be because men no longer felt guilt and remorse for past sins but only regret for past mistakes. They would no longer blame but pity the wrong-doer. They would say 'to know all is to pardon all' and they would pardon all; or rather pardon like forgiveness would disappear. I do not pardon or forgive a stick for tripping me or a man for what he does sleep-walking or hypnotised.

So far we have been discussing theories which argue for a necessary connection between responsibility and determinism. We may now consider a type of theory which makes determination possible though without necessitating it. These theories arise from the possible meanings we might attach to the words 'I could have done otherwise'. First there are obviously cases where the possibilities of alternative action derive from one particular feature of the situation. I went to France last vacation. I could have gone to Chile, so far as time was concerned. But of course I could not go there as I hadn't the necessary money. Or I could have gone to Russia, so far as time and funds were concerned; but I could not get a visa. This type of solution, however, is no help in our present problem. It is not enough that in respect to some one or two conditions I could have done otherwise if it remains the case that in relation to some other feature of the situation I could not have done otherwise.

Another model is that adopted by G. E. Moore. The key here is to regard the phrase 'I could have done otherwise' as a concealed hypothetical. As a parallel to this analysis, take the following examples. 'Jones could easily have won the race. He left his sprint too late' which could reasonably be taken to mean 'Jones would have won the race if he had sprinted sooner'. Or 'Hitler could have won the war. His delay at Dunkirk saved the allies', which again would seem to mean 'Hitler would have won the war if he had continued his drive against Dunkirk'. In both cases it is to be noticed that there is in addition the implication that, *having* delayed his sprint, Jones could not have won the race and, *having* held up his attack, Hitler could not have won the war. The analysis of 'I could have done otherwise' adopted by G. E. Moore is 'I should have done otherwise if I had chosen'. This is clearly a possible meaning of the phrase. It is the sense in which we say 'He was quite free to accept the job but of course he didn't as he dislikes that kind of work'. This means there was nothing to stop him from taking the job except his own desires. But is this an adequate theory to justify praise and blame? It would now seem to depend on the further question which is naturally suggested by the formula 'I should have done otherwise if I had chosen', namely the question 'could I have chosen otherwise?' Moore says that there are two senses in which it may be said that 'I could have chosen otherwise'. First we apply the same analysis and then we have 'I should have chosen otherwise if I had chosen (sc. to choose otherwise)'. He points out that I can force myself to make a choice and sometimes to make a choice which I am reluctant to make, and if I did not make this effort I would choose otherwise. But we are now embarked on a regress. Could I have made the effort I did not make? This would mean 'Should I have made the effort if I had not chosen to do so?' Somewhere we seem driven to decide whether we are simply embarked on a causal series, and if so it is *never* the case that I could have done otherwise *conditions being the same*, but only that I should have done otherwise had they been different. The other sense in which Moore thinks that it can be said that 'I could have chosen otherwise' is that no one could possibly predict which I should choose. This too is a common meaning for 'could' where it is equivalent to 'may' or 'might' expressing ignorance. 'Jones could have been at the party but I didn't see him.' This too, however, is irrelevant to our problem. Whether or not someone knows what I am going to do, the real issue is one of fact not knowledge—the issue whether I could do otherwise in the same circumstances.

Another reason why Moore's analysis is unacceptable is the meaning he attaches to the word 'do'. Take an example. 'I told a lie. I should have told the truth if I had chosen.' Moore sometimes uses the word 'willed' instead of 'chosen'. 'I should have told the truth if I had so willed.' Now what about this distinction between choosing or willing to do something and doing it? In what circumstances could it be said that I chose or willed to do something and did not do it? Perhaps if I were tongue-tied or paralysed. But, as we have seen earlier (p. 53) the judgment of moral praise or blame is passed on what I choose or will and not on what I do (where these diverge). Thus my freedom to do what I will (while of course immensely important in many other respects) is not relevant to questions or moral praise or blame. What matters is my freedom to choose or will.

There are therefore difficulties about accepting any solution which relies on a hypothetical analysis of 'could have done otherwise'. C. D. Broad has put this case effectively. If it can be said that I ought to have acted otherwise, then another action must be *categorically substitutable* for the action I in fact did. Hypothetical substitutability is not enough. Broad finds it impossible to give a clear and satisfactory account of *how* an action can be categorically substitutable for another and concludes that it can never be the case that I ought to have acted otherwise. There are no moral obligations. The other alternative would be that moral obligations exist and that the corollary of categorical substitutability must simply be accepted, as a fact beyond explanation. This may be made more palatable by recalling that explanation would inevitably be in terms of scientific or analytic procedures to which, as I argued above, the self may be recalcitrant.

(Moore himself later agreed that he had been mistaken in thinking that hypothetical substitutability was adequate for moral obligation as a result of criticism by A. C. Garnett, and agreed that it was essential, if obligation is accepted, that a man could have chosen otherwise than he did.)

REFERENCES

The literature on Freewill is vast, and references are given below only to those views and lines of argument examined in the foregoing chapter

Self-determinism:

F. H. Bradley, *Ethical Studies*, ch. i

Responsibility implying Determinism:

R. E. Hobart, article in *Mind* XLIII (1934); P. H. Nowell-Smith, *Ethics*, chs. 19, 20, articles in *Mind* LVII (1948) and LXIII (1954); F. Ebersole, article in *Mind* LXI (1952)

Hypothetical and categorical substitutability:

G. E. Moore, *Ethics*, ch. vi; A. C. Garnett in *The Philosophy of G. E. Moore*, pp. 179–99, accepted by Moore, ibid., pp. 623–4; C. D. Broad, *Determinism, Indeterminism and Libertarianism*, reprinted in *Ethics and the History of Philosophy*; C. L. Stevenson, article in *Mind* (January 1938) reprinted as ch. viii of his *Facts and Values*

11

PUNISHMENT[1]

The problems of punishment are closely connected with those of free will discussed in the previous chapter. The main division between theories of punishment is that which separates the retributive theory from the rest. The retributive theory maintains that punishment is justified because of the wrong done by the person to be punished. The other theories justify punishment by the various consequences which it may produce. Punishment is thus a special case of the dispute between Utilitarians and their opponents which was examined in Chapter 3 of this book. It was there maintained that there are some duties (such as promise-keeping and debt-paying) which depend on previous events, whereas, for the Utilitarian, all duties depend on subsequent events (consequences) for their justification. Is punishment one of these?

The various Utilitarian theories of punishment can be divided into (a) preventive, (b) deterrent, (c) reformatory. When a murderer is hanged or imprisoned for life this *prevents* him from repeating his offence. When a thief is sent to prison this *deters* him and others from stealing again. When a prisoner

[1]. This chapter uses material from an article on Punishment in *Mind* XLVIII (1939) and a discussion note in *Philosophy* XXX (1955) with the permission of the editors of these periodicals

is visited by the chaplain or the psychiatrist or the probation officer and, as a result, decides to 'go straight' in future, this *reforms* him. The deterrent theory is sometimes limited to the effect on other people, but I think this is confusing, because it results in classifying the effect on the criminal as reform. But a criminal who abstains from repeating the offence solely from fear of the punishment is in no way reformed. If he thinks he can get away with it he will do it again. Another possible confusion requires to be cleared away in connection with these Utilitarian theories. A great deal of what is called 'Penal Reform' has nothing to do with the reformatory theory of punishment, and indeed has nothing directly to do with the theory of punishment at all, except in a negative way. The greatest prison reformers have not been concerned with punishment but with its accessories. A sentence of imprisonment need not and should not involve partial starvation, physical maltreatment, or disease. A book which has left a mark on prison administration—*Walls Have Mouths* by W. F. R. Macartney—is full of illustrations of this. His chapter on 'Food' is one, as also is his comment on Governor Clayton's administration.

To keep a man in prison for many years at considerable expense and then free him charged to the eyes with uncontrollable venom and hatred generated by the treatment he has received in gaol, does not appear to be sensible.

Clayton

endeavoured to send a man out of prison in a reasonable state of mind. 'Well, I've done my time. They were not too bad to me. Prison is prison and not a bed of roses. Still they didn't rub it in.'[1]

This reasonable state of mind is one in which a prisoner feels

1. p. 152

he has been punished but not *additionally* maltreated or insulted. We have no more right to keep a convict in a Dartmoor cell 'down which the water trickles night and day'[1] than we have to keep a child in such a place. If our sentimentalists cry 'coddling of prisoners' let them come into the open and incorporate whatever brutality and disease they require into the sentences they propose. Another prisoner, Jim Phelan, makes the point well.

> One of the minor curiosities of jail life was that they quickly provided you with a hundred worries which left you no time or energy for worrying about your sentence, long or short ... Rather as if you were thrown into a fire with spikes in it, and the spikes hurt you so badly you forgot about the fire. But then your punishment would *be* the spikes, not the fire. Why did they pretend it was only the fire when they knew very well about the spikes?[2]

It may be thought odd to go to criminals for arguments on punishment. But, after all, they are the people whom we are supposed (on these Utilitarian theories) to be reforming and deterring; and what they think of our doings is of primary significance. But they are not the only authorities who make the point. The greatest recent authority, Sir Alexander Paterson, says

> The first duty of a prison as an institution of the State is to perform the function assigned to it by the law; and its administration must ensure that a sentence of imprisonment is a form of punishment. It must be clear at the outset, however, that it is the sentence of imprisonment which is the punishment and not the treatment accorded in prison. Men are sent to prison *as* a punishment and not *for* punishment. It is doubtful whether any of the amenities granted

---

1. Ibid., p. 258
2. *Lifer*, p. 40

in some modern prisons can in any way compensate for the punishment involved in deprivation of liberty.[1]

The last sentence is over-cautious. Let any 'amenity prison' offer release to its inmates and how many would stay to enjoy its amenities? But this in turn suggests that Paterson himself is not quite clear. A prison administration need not 'ensure that a sentence of imprisonment is a form of punishment'. All it need do is to ensure that its prisoners do not escape, and otherwise that they suffer no 'uncovenanted' maltreatment.

Having cleared away these possible confusions, we may now turn to punishment itself. The reform theory involves, in some cases, the confusion we noted earlier between punishment and its accessories. The visit of the prison chaplain (or the string quartette) are not *parts* of the punishment, nor is a criminal imprisoned in order that he may be so visited. The duty of reforming prisoners is a different duty from that of punishing them. A parallel is the case of tact and truth. If you have to tell someone an unpleasant truth you may do what you can to spare his feelings and soften the blow while still making sure that he understands your meaning. Here no one would say that your care of him before and after are *reasons* for telling him the truth. You do not tell him the truth in order to spare his feelings; but having to tell him the truth you try also to spare his feelings. So prison authorities may make it possible for a criminal to reform. They cannot ensure this, and, if they fail, the punishment would be no less justified.

Some moralists see this and exclude such 'extra' arrangements for reform from their theory of punishment. They say the punishment *itself* must reform. It should bring home to the criminal the error of his ways. It should teach him a salutary lesson. Now it is certainly a good thing if these results occur, and once again special efforts may be made, by the judge when he

---

1. *Paterson on Prisons*, p. 23

pronounces his sentence, for example, to try to achieve them. But once again if they meet with bravado or bitterness and fail, this does not show that the punishment has lost any of its justification. Reform, therefore, while it seems to be an admirable by-product of punishment, does not appear to constitute its justification.

We are thus left with prevention and deterrence on the one hand and retribution on the other. It is sometimes urged against the deterrence theory that it would justify stepping up punishments to a maximum. So long as *anyone* breaks a given law this shows the punishment was not enough. There are two objections to this conclusion. First, it is said that such severity would defeat its own object. When the penalty for sheep-stealing was death, juries refused to convict even those thieves caught in the act. This argument might be met by abolishing the jury system and leaving it to Judge Jeffreys to pass both verdict and sentence. And, in any case, the objection is itself a retributivist one, for the jury say that death is *too heavy* a sentence for sheep-stealing (and 'too heavy' does *not* mean 'unnecessary for deterrent reasons'). The other objection to maximum severity goes further to meet it. It is that if minor crimes are severely punished major crimes will be encouraged. If a man is to be hanged for sheep-stealing he will shoot his way out of trouble. Thus over-enthusiasm for deterring sheep-stealers encourages murder. But even this argument does not go the whole way. As long as there can be *some* more severe sentence for the possible added crime the lesser one can still be punished severely. So this remains a difficulty.

The retributive theory has very few defenders in these days. Philosophers have mostly rejected it and practical men have welcomed its decline in our penal arrangements. Yet some of the arguments against it are certainly answerable. Rashdall associates retributive punishment with vengeance and even calls it 'the vindictive theory'.[1] He argues that revenge is an

1. *The Theory of Good and Evil*, ch. ix

immoral attitude and that retribution is incompatible with forgiveness which is the appropriate response to injury. Now all this is confusion of terms. The only person who can feel vengeful or vindictive where an injury is done is the injured party. The name of the feeling of a third party on such an occasion is 'indignation'. While vindictiveness is certainly an unchristian and immoral attitude, though a very natural one, moral indignation is surely entirely respectable. So far then as punishment is not inflicted by the injured party it cannot express vindictiveness, though it can express indignation. Similarly with forgiveness. The only person who can and therefore ought to forgive is the injured person. If A damages B, it is absurd for C to forgive A. The appropriate word here is 'condone'. Thus there is no inconsistency between forgiveness and retributive punishment since these duties fall on different shoulders. It is true that this conclusion is sometimes avoided by maintaining that 'society' is the injured party and therefore 'society' is taking revenge on the criminal. But here, as elsewhere, the treatment of society as a moral agent leads to error and confusion. Statements about the moral feelings and actions of 'society' have to be analysed in terms of the feelings of individuals and the actions of citizens and officials. And in the present case 'society feels vindictive' has to be analysed as 'citizens feel indignant'.

Rashdall also describes retributive punishment as 'the infliction of pain for pain's sake'[1] and as 'adding evil to evil'. The first objection to this is that pain is not evil. If a man has a toothache in bed, is there something evil in his bed? Pain is in some sense bad, regrettable, to be removed if possible; but the dentist is not a moral straightener. Then, secondly, pain is no part of punishment. Only flogging, among all our penalties, is the infliction of pain and it is on the way out. But, says Rashdall, the pain may be mental or physical. Here again

1. Ibid., p. 286

surely is an abuse of language. Our standard punishments are not inflictions of suffering but deprivations: of life (capital punishment) of liberty (imprisonment) and of property (fines). It is true that no one *likes* to be deprived of any of these good things. But to represent these dislikes as pains or evils is a mistake. It is curious that Rashdall did not revise the text of his argument for he admits in a footnote that

Pain is an accident of retribution and I am not aware that I made it even an inseparable accident. If a criminal is shot, are we, if there is no pain, to say that there is no retribution?[1]

The Royal Commission on Capital Punishment defends hanging as the most painless method of execution.[2] If it be said that this does not diminish the mental agony of waiting for the execution, the answer is that the aim of the sentence is not the infliction of this agony. If it were, the duration would be specified and the suffering would be an essential part of the punishment. The Commission argue[3] that the delay is regrettable, but is to be defended as allowing time for an appeal or application for reprieve. And 'the preliminaries of the execution should be free from anything that unnecessarily sharpens the prisoner's apprehension'.[4] The world is a worse place the more evil there is in it, and the more unnecessary suffering there is in it. But it does not seem to me necessarily a worse place whenever men are deprived of something they would like to retain. And this is the essence of modern punishment.

A preliminary difficulty about the retributive theory is raised by the question 'What is the retribution for?' Most holders of the theory have answered this by saying 'for moral guilt', though some would restrict this to anti-social wrong-

1. Ibid., p. 287
2. *Report*, paras. 726–31
3. Ibid., para. 763
4. Ibid., para. 724

doing. My difficulty with this view is the question of status. It takes two to make a punishment, and for a moral or social offence I can find no punisher. I am therefore inclined to think that punishment is retribution for *crimes* not for *sin*. A criminal is a man who has broken the law. Many bad men are not criminals. An 'innocent man' is a man who has not broken the law, in connection with which he is being tried, though he may be a bad man and have broken other laws. We may be tempted to say when we hear of some brutal assault 'he ought to be punished' but I cannot see how there can be duties which are nobody's duties. If I see a man beating a horse in a country where there is no law against cruelty to animals, I cannot say 'I am now going to punish you'. He will reply rightly 'Who are you?' I may have a duty to try to stop him and one way may be to hit him or another may be to buy the horse. Neither the blow nor the price is a punishment. For moral offences, God alone has the status necessary to punish: and the theologians are far from agreement about whether this status is to be brought into action. There is of course much confusion about this. Many people think of all kinds of suffering as being visited on men 'for their sins'. In Dostoievsky's *Crime and Punishment* Raskolnikov found he was losing control of his thoughts after the murder and cried 'Can this be the punishment already beginning? Indeed, indeed, it is!' Here as usual in Dostoievsky the whole moral atmosphere is deeply theological. The police come in late and almost as an irrelevance.

It seems to me then that punishment is a corollary of lawbreaking. The crucial difficulty about the deterrence theory (and this applies to the reform theory too) is that it would justify the punishment of an innocent man—that is to say a man has not committed the crime for which he was punished —provided he was believed to be guilty by those likely to commit the crime in future (or, in the case of the reform theory, provided that his treatment in prison improved his character). A typical illustration is that of Oscar Slater. He was prosecuted, on very unsatisfactory evidence, for the murder of Mrs Gilchrist in 1908. Public opinion was violently against him and he was sentenced to death. If he had been executed, the deterrent effect on other would-be murderers would have been exactly the same as if he had been guilty. The Secretary for Scotland, however, against the advice of his legal advisors, reprieved Slater; and, after twenty years of agitation by legal experts, his conviction was 'set aside' and he was consoled by a grant of £6,000. Only one fact can justify a man's punishment and that is a *past* fact, that he has broken a law.

Macartney confirms this line. It is striking that he never uses the word 'injustice' to describe the brutality and provocation he experienced. In his view only two types of prisoner were *unjustly* imprisoned, those who were insane and not responsible for the acts for which they were punished,[1] and those who had not broken the law.[2] It is irrelevant that some of these latter were, like Steinie Morrison, dangerous and violent characters who, on utilitarian grounds, were well out of the way. That made their punishment no whit less unjust.[3] And, for a specific instance, he cites the sentences on the Dartmoor mutineers.

The Penal Servitude Act . . . lays down specific punishments for mutiny and incitement to mutiny, which include flogging . . . Yet on the occasion of the only big mutiny in an English prison, men are not dealt with by the Act specially passed to meet mutiny in prison . . . but under an Act expressly passed to curb and curtail the Chartists—a revolutionary movement.[4]

Here again the injustice lies in condemning men for breaking a law they did not break.

1. Macartney, *Walls Have Mouths*, pp. 165–6
2. Ibid., p. 298
3. Ibid., p. 301
4. Ibid., p. 255

Hegel argues that retributive punishment alone honours the criminal (in contrast with the view that deterrence and reform are the modern, progressive, reasonable theories). For it treats him as an adult rational being with a choice of conduct. Deterrence and reform treat him either as a means to frighten other people or as requiring remoulding by fear or treatment. Macartney is with Hegel here: 'To punish a man is to treat him as an equal. To be punished for an offence against rules is a sane man's right'.[1]

An attempt has been made by A. M. Quinton[2] to meet this demand, that nothing but law-breaking can justify punishment, by treating this as a logical or linguistic point. The *meaning* of 'punishment' includes a reference to previous crime. This means that it is *logically* impossible to punish the innocent. We *can* only punish the guilty. But this does not mean that we ought to punish them. The reasons for this are the utilitarian reasons of deterrence and reform.

Now it seems to me that Quinton is right on the linguistic or logical issue. The meaning of 'punishment' does include a reference to the past. But this does not exclude an ethical judgment on the procedure so described. 'Divorce', 'revenge', and 'forgiveness' all include in their meaning a reference to the past. But this does not prevent us from condemning revenge, commending forgiveness, and differing about the morality of divorce.

I maintain that the officer of a society whose rule has been broken not only *can* but ought to punish the offender. Quinton recognises that this is a possible account of one *kind* of legal system, a system in which *fixed* penalties are allocated for breaches of rules. I take it that he would apply this to those cases in actual systems for which fixed penalties are laid down. So an English judge not only can but has a duty to impose the

1. Ibid., p. 165
2. *Analysis*, vol. 14, p. 133

death sentence on a person found guilty of murder. I assume Quinton would not hold in this case that he *can* impose the death sentence but ought to impose whatever sentence will have the best consequences. Quinton points out that, in cases other than murder, the penalty laid down is a maximum and the judge has discretion within it. So here he should decide the actual sentence by considerations of utility. To which I would give two answers. First, the discretion is given by the law; so, even if it would take more than forty shillings to deter or fourteen days to reform, the judge has no right to impose sentences above these maxima. Secondly, I do not admit that the considerations which fix the penalty within the maximum are utilitarian. They normally concern not the effect on the criminal or others in the future, but the degree of guilt in the past, or responsibility in the past also. Negligence varies in degree; one man knew his brakes were faulty and did nothing about it, the other did not know, though he could and should have found out. Conspirators vary in responsibility; one is ring-leader, another his tool. The Report of the Royal Commission on Capital Punishment says

Offences of the same legal category differ greatly in gravity and turpitude and the courts make full use of the wide range of penalties they have power to impose.[1]

Moreover the vast majority of sentence-revisions by Courts of Appeal take the form 'The sentence of . . . appears to us too heavy (or inadequate) for the offence of . . .'.

The line adopted above takes the existence of laws for granted, and it is at this point that utilitarian considerations are relevant. They are considerations for legislators. Should there be laws, and what laws should they be? Should penalties be attached to them and if so what penalties, and should they be fixed or maxima? None of these choices is Hobson's choice;

[1]. *Report*, para. 20

and the majority of them are Utilitarian. The association of a penalty with a law is certainly deterrent. Legislators make a choice here. They do not choose to punish. They hope the threat will succeed and no punishments will be needed. Their laws would thus have the best results possible. Many men obey the law because they see its order is reasonable, some because they trust the authorities, some from inertia, some from fear. Over this whole field, and it will include the majority of the citizens, law achieves its ends without punishment. Punishment then is not a mere corollary of law. Another choice beside that of the legislators is needed; and that is the criminal's choice. He 'brings it on himself'. Here, as often, the sound commonsense of Dr Johnson outpaced the wits of many cleverer men. He said to Boswell:

> To punish fraud when it is detected is the proper art of vindictive justice; but to prevent fraud and make punishment unnecessary is the great employment of legislative wisdom.

Note. The above extended discussion of punishment has been included because it illustrates a number of general points raised in previous chapters: the rejection of Utilitarianism, the distinction between justifying a particular act and justifying a rule, the justification of rules by their effects, the distinction between linguistic and moral issues, and the moral relevance of free choice.

REFERENCES

(apart from those given in the text of the chapter)

A. C. Ewing, *The Morality of Punishment*
G. W. F. Hegel, *Philosophy of Right* (tr. T. M. Knox), pp. 66–74
F. H. Bradley, *Ethical Studies*, ch. i

12

RIGHTS AND DUTIES

There is one sense of the word 'right' which we have not yet examined. Previously we considered its use to describe the action appropriate to the circumstances in which a moral agent finds himself. But there is, besides this mandatory or obligatory use, a *permissive* use. 'Will it be all right for me to go now?' 'Yes certainly, but stay if you like.' Connected with this use, but not identical with it, is the use when a man is said to have *a right* to something. Connected, because he need not exercise his rights. Just as it is all right to go but he may stay, so I have a right to free speech even if I remain silent. But the uses are not identical; for the assertion that I have rights seems to involve claims on other people, though just what these are will be a problem for later discussion. The purely permissive use of 'all right' seems to be basically negative. I shall not do wrong if I go (nor if I stay). There is no objection to my going. The objection need not be a moral one. An employee may be told that it is all right for him to go; his employer does not need him any longer tonight.

What is meant by having a right? Rights are both positive and negative. I have a right to free speech or to the practice of my religion; this means that I should be left free from interference in these fields. Or I may say I have a right to a living wage or to medical treatment; this means that these things should be provided for me. It now seems clear in what sense rights are claims on other people. When I have a right, someone else, or other people, or the authorities have a duty to leave me alone or to provide the service I claim. Rights imply duties. Is the reverse the case? Take the following definition of a right:

A man has a right whenever other men ought not to prevent him doing what he wants or refuse him some service he asks for or needs.[1]

The word 'whenever' implies that if A has a duty to leave B alone or to provide him with some service, then B has a right to that permission or service. The relation between rights and duties is then one which works both ways. Now I find this doubtful. If I am asked by a colleague for the loan of a book which he needs and if I am not using it, I think I have a duty to lend it to him; but I do not think he has a right to it. If I have a seat in my car and I overtake a weary walker I think I ought to give him a lift; but I do not think he has a right to it.

Of course it is possible to avoid this difficulty by restricting the use of the word 'duty' so that it applies only to those obligations to which others have rights. One's duties could then be a minimal set of moral obligations, the discharge of which would still leave one free and indeed obliged to go on to a higher level of moral actions. We shall return to these two levels later.

But first there are some other points about rights to be cleared up. A distinction must be drawn between legal and moral rights. The actions in which the law protects me and the services and privileges the law awards me are my legal rights.

1. J. P. Plamenatz, *Proceedings of the Aristotelian Society*, suppl. vol. XXIV, p. 75

When I go to my solicitor and ask him what are my rights in a certain situation, legal rights are in question. But we require, in addition, the notion of moral rights. For women and slaves and negroes have all claimed rights (or had rights claimed for them) before they had corresponding legal rights.

This leads to the question how far rights depend on society and on social recognition. Legal rights obviously do so depend. But it seems impossible to say that moral rights depend on social recognition. Slaves and women had rights before these were socially recognised. A right need not even be recognised by the person who has it. The definition above, in saying a service must be 'asked for or needed', seems to fall into this error. Children have a right to education even though they do not know this and therefore cannot ask for it. Of course in one sense rights may be said to be social and to require recognition. If it is true that when a man has rights other people have duties, then in this sense rights involve social relations. And if anyone is to *assert* that a man has a right then at least the speaker has to recognise the right. But the 'social relations' could be limited to a couple of people; and it would seem reasonable to hold that slaves had rights even when no one asserted this and consequently no one recognised it.

We now return to the problem of distinguishing those moral obligations to whose performance other people have a right from those to which they have not; or (on the restricted view of duty) of distinguishing duties from other moral obligations. J. S. Mill draws the latter distinction but does not explain it very satisfactorily.

> Duty is a thing that may be *exacted* from a person, as one exacts a debt... Justice implies something which it is not only right to do and wrong not to do, but which some individual person can claim from us as his moral right. No one has a moral right to our generosity or benevolence, because we are not morally bound to practise these virtues towards any given individual... If a moralist attempts

to make out that mankind generally though not any given individual have a right to all the good we can do them, he at once by that thesis includes generosity and beneficence within the category of justice. He is obliged to say that our utmost exertions are *due* to our fellow creatures, thus assimilating them to a debt.[1]

Mill seems here to think that *duties* are always owed to specific individuals whereas generosity and beneficence are not. It is true that in some cases particular individuals have rights against me (my creditor, my employer). But the negative rights (not to be interfered with) are not claims on particular people but on people in general. Mill thus provides no criterion for distinguishing duties and justice from moral obligations which go beyond these. These higher level obligations were called by the medieval philosophers 'duties of supererogation'. But they too failed to provide a very convincing distinguishing feature for them. In St Thomas Aquinas, 'supererogatory' has two meanings: (a) when a certain act is not necessary to salvation, (b) when an act is necessary to salvation, but there is a choice between alternative methods of doing it. The second sense is clearly not relevant here. The first does not seem clear by itself. But if we ask what Aquinas meant by being deprived of salvation, it may be said that this is equivalent to holding that neglect of duties is punishable, but neglect of acts of supererogation is not. Mill comes close to this answer too.

We call any conduct wrong according as we think the person ought to be punished for it. And we say it would be right to do so and so *or* merely that it would be desirable or laudable according as we should wish to see the person whom it concerns compelled *or* only persuaded and exhorted to act in that manner.

It might be thought that this reference to punishment is too narrow (though we shall see grounds for accepting it later in

1. *Utilitarianism* (Everyman ed.), ch. v, pp. 45–7

Rights and duties

this argument). If so it might suffice to say that people should be blamed for not doing duties but not for not doing supererogatory actions. We might be tempted to add, with an eye on symmetry, that people are not praised for doing duties but are praised for doing supererogatory actions. We do not praise judges for passing the sentence prescribed by law, nor debtors for paying their debts nor our friends for keeping their dates with us, nor the passerby for telling us truthfully the way to the station. All these are certainly duties, which it would be blameworthy *not* to perform. But the second point is certainly mistaken. We do sometimes praise men for doing one of these duties. It depends how difficult the duty is. When Regulus was sent by his Carthaginian captors to Rome to try to negotiate a peace he promised to return. The Romans pressed him hard not to go back but to lead them in continuing the war. His decision to return was surely praiseworthy. We should also be inclined to praise Sir Walter Scott's long struggle to repay his debts. And even the first half of the distinction is doubtful—that men are not to be blamed for not doing supererogatory duties. Men certainly feel an obligation to do supererogatory duties and blame themselves when they fail to do them. 'We have not done what we ought to have done.' If freedom from blame goes with failing in these duties then the Saints clearly would say that they had *no* supererogatory duties.

A different line to distinguish duties which involve rights from those which do not may now be considered. It is suggested in a remark by Tom Paine: 'Whatever is my right as a man is also the right of another and it becomes my duty to *guarantee* as well as to possess.'[1] These rights should be *guaranteed*. This means that my enjoyment of negative rights (free speech, religious freedom) should be *protected* from those who have a duty to respect them but do not do so. And that my enjoyment of positive rights should be *ensured by compulsion* if need be.

1. *The Rights of Man* (Everyman ed.), p. 98

We can amend the definition given at the beginning of this chapter (p. 140) to read: 'A man has a right whenever other people should be *stopped* from preventing him from doing something or *compelled* to provide him with some service.'

There is no reference here to the identity of the people who should prevent or compel. Tom Paine's view that *I* myself should guarantee to others the rights I claim for myself makes an impossible demand on the individual. The clue is given, however, in his first sentence. A right I have as a man is a right all men have. And, even if there are restricted classes of claimants, the claim concerns classes of people, not individuals. If a child has a right to education, all children have a right to education. If an able child has a right to university education, then all able children have this right. Now the only body which can *guarantee* rights to all men or all children or even all gifted children is an inclusive body with compulsory powers. The State comes nearest these requirements. But there are some rights which even the State cannot guarantee—for example the repayment of debts by foreigners. And, while there are some rights the State could guarantee, it may fail to do this through poverty or policy. Then what? The final resort would seem to be international action, where an international body would guarantee those rights which States cannot or will not guarantee. The work of the League of Nations against slavery and the Charter of Human Rights of the United Nations are both attempts to put pressure on States which do not guarantee the rights of men. Here we come back to Mill's view, that a wrong is something for which a person ought to be punished, and a right something he should be compelled to respect.

The only recent philosopher who has emphasised the difference between these two moralities, the morality of rights and claims and the morality of supererogation, is Bergson. The first morality he says is social, the second human.

Our duties are determined by our relations to particular individuals within the associations to which we belong. The obligations of this first morality are prescribed by rules and correlated with rights. We feel these obligations as a pressure and a constraint. The higher morality involves duties which, if owed to anyone, are owed to mankind. This morality comes before us not as a code of rules but through a living individual example. It is therefore frequently associated with religion. But this is not a necessary connection, for anyone may have a hero and try to live up to his example. The obligations here concerned do not affect us by external pressure and constraint but by attraction and magnetic appeal.

A good half of our morality comprises duties whose obligatory character is explained in the first analysis by the pressure of society on the individual; we accept it without too much trouble because these duties are regularly practised, because they are clear and definite, and because it is easy for us to grasp their fully visible part and descend thence to their roots and discover the social requirements from which they take their rise. But that the other half of morality is the translation of an emotion in which one yields not to a pressure but to an attraction many would hesitate to admit. The reason is that it is not possible in most cases to discover in the depths of one's own personality the original emotion. It has left behind it certain formulae in which are enshrined in the social conscience what was originally immanent in that emotion—a new conception of life or rather a new attitude towards it.[1]

Bergson describes this distinction as one between an *open* and a *closed* morality. He criticises convincingly the suggestion that the move from particular obligations within a closed society to obligations to human beings as such is simply a *widening* like that from family to tribe or tribe to nation. The widest service differs in kind and not in degree only from the service of family or fellow-countrymen. These services are exclusive

1. *Les Deux Sources de la Morale et de la Religion*, p. 46

and consequently combine naturally and easily with distrust and hostility towards outsiders. They are also 'natural' to man. The Christian love which includes loving one's enemies is not a natural feeling but something which can come only with a new vision.

Bergson also explains convincingly how it is that the frontier between the two moralities seems to shift and become blurred. The original inspiration of the saint or hero fades, leaving only his 'message'. This message in turn crystallises into new rules which look just like the old social duties of the closed society, and borrow from them an obligatory character.

> We find ourselves before the ashes of a dead emotion and because the driving force of that emotion came from the fire within it, the formulae which survive it would be unable to arouse our wills if the more ancient rules which express the basic needs of social life did not communicate to them by infection some of their obligatory character. Then the two juxtaposed moralities seem to blend, the second derives an obligatory character from the first, the first a widened application from the second. But remove the ashes and you will find heat still below and the spark can be rekindled, the fire relit, and it will then spread from point to point.[1]

Thus justice is enlarged by charity and what was once thought to be charity comes more and more to seem simple justice. So we have to make the distinction, but it is continually being lost and won again. Hence there can be no criterion for distinguishing those obligations which are supererogatory from the rest. For with each age the frontier moves. Objects of charity—children, the old, the unemployed—become bearers of rights.

We may now return to the use of 'right' in such phrases as 'It is all right for you to go now'. This, we suggested, marks a choice as morally neutral. It would not be wrong for you to go

1. Loc. cit.

(or to stay). This in turn leads to the question of distinguishing moral from non-moral decisions and the area of moral conduct from that of non-moral conduct.

Many people, if asked how many moral decisions they made on a given day, or how long it is since they made a moral decision, might well answer 'very few' or 'a long time'. This is because it is natural to limit 'moral decision' to cases in which two alternatives are explicitly present to the mind—either two alternative moral courses or one moral and one immoral—and where there is a conscious and explicit moral judgment. This restriction is also influenced by the tendency to think that moral action must be preceded by moral judgment. This emphasis on moral judgment is mistaken. We seldom have to make such judgments; normally only when we are uncertain of what is right or where we have to educate or judge other people. But most of us spend little time in moral indecision and still less in preaching or writing testimonials. The primary job of a moral agent is moral action, and only in a few cases are moral actions themselves the making of judgments or require to be preceded by such judgments. (The error here is a particular case of the general error of thinking that intelligent action of any kind is action preceded by thinking. This error, with Professor Ryle's criticism of it, was noticed previously in Chapter 6.) We may grant that actions are moral if they are the execution of moral judgments or decisions and that the making of moral judgments is itself a type of moral action. But how are we to decide whether an action is moral when no explicit moral judgment or decision seems to precede or accompany it? Some of them certainly are but some certainly are not. Even if we might find it difficult to say of any field of action *as a whole* (e.g. hobbies or other leisure activities) that it was morally indifferent or neutral what we did or how we did it, we do believe that many particular choices we make are indifferent or neutral, for example whether I pay a debt by cheque or cash or tell a man the way by talking or pointing. If it is objected that these are trifling matters then it is possible to find decisions which make a considerable difference to my life and are yet not moral decisions: whether to give up skiing altogether, where to go on sabbatical leave, whether to lecture next term or the term after. The common answer here is that any of these decisions *may* be a moral decision. If my continued skiing involves leaving my wife at home, if I am not very fit and Yorkshire air will do me more good on my leave than Venice, if lectures on my subject are short next term—well then, of course, the case is altered. But if my wife would welcome my absence to get the house redecorated, or if I am fit and would find Venice no menace, or if lectures are in equally good supply both terms—then it does not make a particular choice moral to say that in different circumstances a similar choice would be a moral choice.

But it is not always so easy to be sure whether actions are moral when no conscious moral judgments or explicit moral decisions precede or accompany them. Take a morning's actions. I write letters from 9 to 10, I teach from 10 to 12, I attend a committee from 12 to 1. How much of all this is moral action and how much indubitably neutral? Some perhaps counts as moral. My 11 o'clock pupil is both stupid and obstinate and I hope I show what an impartial observer would consider to be patience and tolerance. I hurry to my committee so as not to keep the others waiting. But what of all the rest, the steady doing of the job (with the 10 o'clock pupils who are neither stupid nor obstinate), my contributions to the discussions of the committee? I should be astonished to be praised (or blamed) for taking up these rather than those points in my pupils' essays or for the points I made in the committee discussion.

To distinguish the moral actions from the others we should, at any rate, include the cases in which, if questioned, I should have formulated a moral judgment as, at least, part of the explanation of my action, as its motive or intention. I may never say to myself as I set out or go to the meeting 'I ought to hurry' or 'it would be wrong to keep them waiting'. But nevertheless if I had been stopped, in the street, by someone who said 'Why the hurry?' my answer, and an honest answer, would have been 'because I mustn't keep them waiting' and not 'because I am trying to get warm' or 'for exercise'. If either of the latter explanations had been the true one, I should have said my hurrying was non-moral and of no concern to moral philosophy. This of course is not always so easy. My explanation may not be honest, or I may not really know why I am hurrying or my motive may be mixed. (If it was a hot day, I should not be going so fast.) But there may be good evidence here all the same. There is my relief when I arrived at the meeting and found all the others there but the clock at 11.58, or the clock at 12.2 but two other members missing. As to mixed motives, I must ask myself 'Should I have hurried even if I had not been cold?' This may not be easy to answer though again there may be evidence that I hurried last time when it was warm, and that I continued to hurry this time even after I got warm. But even if decision is difficult in a particular case this does not matter if the principle is clear. If I do not know whether I can honestly say 'I hurried so as not to keep them waiting', then I cannot say whether my act was moral. Moral philosophy cannot decide whether particular acts are moral but only under what conditions an act would be moral.

Even the inclusion of actions which would be explained by a moral reason still leaves many, perhaps most, everyday actions as neutral. We still ask what we are to say about the steady doing of the job, the normal contributions to a committee's discussion, the weekend's leisure. The word 'steady' perhaps begs the question by suggesting the overcoming of occasional reluctances or difficulties; for these occasions would probably be consciously moral choices. 'Routine' would be better, provided that it does not imply mere habitual repetition, which in many jobs (teaching or medicine) would impair the performance. In most jobs, however, there are general considerations which must be regarded as moral. There is earning one's wage, and not letting down one's fellow-workers, besides (in some fortunate cases) the worth-while character of the job itself. How far are such motives operative in everyday life? We all know cases where they are conspicuously absent, cases where the placard 'Men at work' is an irony. But the difficulty arises when people seem to be just jogging along, not conspicuously slacking nor conspicuously stakhanovite. It arises even more acutely when a man obviously enjoys his work, like the artist painting a picture or the professional cricketer knocking up a century. As before, we must say that it may be impossible to tell whether any such activity has a moral element in it. As with mixed motives, we may not know whether the moral factor is there, nor if it is in what strength. But again there may be some evidence. The word 'steady', for example, provides it. If the joy in the work flags, if difficulties come, but yet the work goes on, this is evidence that the moral factor was there all the time, though it needed an obstacle to reveal it. In the distance the river may look still and motionless, but where a rock sticks up the strength of the current is revealed. There are also usually some parts of any job which are not likely to be enjoyed. There are the administrative details of teaching and doctoring, the recognition of the work of others (colleagues or subordinates), and if these are well handled, the moral element is clear. So too with the committee; the matter under discussion may be in itself interesting or a man may just enjoy discussion. But it is still possible to notice whether a man slumps back in the chair as soon as the item which interests him is settled, how fair he is to points made against him, how far he is prepared to revise his own view when the point against him is a good one, how far he raises difficulties just for the fun of discussing them, and so on. And so too again with leisure. Even though a man is doing just what he likes to do with his Saturday afternoon, there are still ways and ways of doing it which in most cases would give some clue as to whether he is wholly self-centred. There are not here, of course, as there are with the job (and perhaps the committee) the overall moral principles controlling the situation: value for money, the value of the work itself, the concern for colleagues and subordinates. Leisure is to that extent a moral holiday too. But there remain the claims of one's family, the general public, one's fellow-players—all of them very relevant, for example, to the Bank Holiday motorist.

So after all the field of morality is not so limited as at first seemed likely, though it must still be emphasised that there are wholly non-moral choices and wholly non-moral factors in choices with mixed motives.

13

CONCLUSION

There are two lines of criticism which might very properly be urged against the preceding chapters. First that the examples are very uninteresting and second that very different types of problem are not adequately distinguished.

The choice of examples has been an intentional policy. It would have been much more interesting to look for examples of heroic virtue, or of really difficult complexity, as the existentialists do. Keeping promises, paying debts, telling the truth—these are the small change of moral lives. The two difficulties about the more fascinating examples are these. First they require a great deal more space to set them out; and secondly they are liable to deflect attention from the point they are illustrating especially if (as would usually be the case) they are themselves debated issues. The middle of the road is not an exciting area and one meets few fascinating creatures there in comparison with the gutter and the ditch, but one gets along faster and with fewer distractions.

The confusion of types of problem is a more serious matter. There are first the most general and abstract questions; the distinction between means and ends; between factual and moral beliefs; between description and expression. Then there are questions like the issue between the Utilitarians and their opponents, between free will and determinism, between subjectivism and objectivism. The first range of questions could have been fully discussed without raising or deciding any of the second range. Then thirdly there are the questions about what *has* moral value or other value. This is the issue between the pleasure-Utilitarians and G. E. Moore. And finally there is the discussion of the rights and wrongs of some particular kind of action. In this book, the chapter on punishment might be held to fall under this head.

One reason why these distinctions have not been drawn earlier and then adopted as the basis of this book is that those who draw them tend to exclude some of these areas from moral philosophy altogether. I have not wished to do this because I am not convinced by these restrictionist policies. It is arguable that a moral philosopher should not discuss the rights and wrongs of particular moral issues; and there are people who have done this effectively who are not philosophers—Shaw and Ibsen on marriage and divorce for example. But otherwise I see no reason for excluding any of the areas distinguished above. The reason for not separating them was that the theories I wished to discuss did not separate them. If they had been separated the problem of order of discussion would have been insuperable. To start with the most abstract section (what has been called meta-ethics) would have involved a very difficult argument and topics including hardly any specifically *moral* philosophy. To begin with the later sections would have raised questions from the abstract section without answering them. Hence I think the drawing and justifying of these distinctions is a task to be faced late in the day and not one to be done first as the basis for the rest of the argument.

INDEX

ANALYTIC Propositions, 94–5
Appeal, Courts of, 137
Aquinas, St Thomas, 142
Attitudes, moral, 82–4
Ayer, A. J., 77–9, 82–5, 91, 95, 108

BAIER, K., 64
Barnes, W. H. F., 102, 105
Benn, S. I. and Peters, R. S., 50
Bentham, J., 17–21
Bergson, H., 145–6
Berkeley, G., 98
Black, M., 73
Bradley, F. H., 87, 118, 138
Broad, C. D., 108, 126
Burke, E., 41
Butler, J., 17

CAPITAL Punishment, Royal Commission Report on, 133, 137
Carritt, E. F., 57–9, 63, 70
Chamberlain, N., 24
Charter of Human Rights, 144
Churchill, W. S., 95–6, 119

Couéism, 89
Croce, B., 114

DARWIN, C., 110
Determinism, ch. 10 *passim*
Deterrence in punishment, ch. 11 *passim*
Dostoievsky, F., 134
Duelling, 41–2

EBERSOLE, F., 120–3, 126
Ellis, W. W. 43
Emotive theory, ch. 8 *passim*
Ewing, A. C., 138

FLEW, A. G. N., 73
Foot, P., 73
Ford, F. M., 55
France, A., 100, 105
Free will, ch. 10 *passim*
Freud, S., 110

GALILEO, G., 89
Gallie, I., 42, 64

Garnett, A. C., 126
Goebbels, J., 15-16
Graham, M., 114

HAMLET, 64
Hare, R. M., 10
Harrison, J., 50
Hedonistic Utilitarianism, ch. 2 *passim*
Hegel, G. W. F., 41, 136, 138
Heidegger, M., 101
Hitler, A., 95-6, 124
Hobart, R. E., 119-20, 126
Hobbes, T., 75
Horsburgh, H. J. N., 64
Housman, A. E., 81, 115
Hughes, G. E., 64
Hume, D., 72, 75-6, 79
Hutcheson, F., 40

JOHNSON, S., 138
Joseph, H. W. B., 49
Judgment, faculty of, 106-7

KANT, I., 28, 38-40
Kierkegaard, S., 101, 113

LANGFORD, C. H., 108
Lawrence, T. E., 114
League of Nations, 144
Lewis, H. D., 64

MACARTNEY, W. F. R., 126, 135-6
Macdonald, M., 102-3
McLellan, J. E., 73
Marcel, G., 101
Marx, K., 110, 116

Mayo, B., 10
Mill, J. S., 10, 17-21, 49-50, 141-2, 144
Moore, G. E., 10, 21, ch. 3 *passim*, 37-8, 41, 47-8, 51-2, 56, 70, 85-6, 89, 107, 124-6, 153
Moral and non-moral actions, 13-14, 90-1, 147-51
Motives and duty, 21, 23-4, 59-63
Munich Agreement, 23-4

NARVESON, J., 31
Nowell-Smith, P. H., 10, 26, 120-1, 126
Nuclear Disarmament, 80
Nystedt, H., 64

OBJECTIVISM (of duty), ch. 5 *passim*
Objectivism (of moral values generally), ch. 9 *passim*
Organic unities, 22
Ought implies can, 60-3, 109, 118-26

PACIFISM, 25, 29
Pain and pleasure, 20
Pain and punishment, 132-3
Paine, T., 143
Paterson, A., 129-30
Pavlov, I., 110
Penal Reform, 128
Persuasive judgments, 85-90
Phelan, J., 129
Phillips, D., 73
Picasso, P., 99
Plamenatz, J. P., 140
Plato, 9, 45

Index

Pleasure, ch. 2 *passim*
Predictability of actions, 110–17
Prichard, H. A., 25–6, 31, ch. 5 *passim*, 67–8, 70–1
Punishment, 120–3, ch. 11 *passim*

QUINTON, A. M., 136–7

RASHDALL, H., 10, 132
Rawls, J., 49, 50
Reformation in punishment, ch. 11 *passim*
Retribution in punishment, ch. 11 *passim*
Rights, ch. 12 *passim*
Ross, W. D., 25–30, 31–3, 48, 51–6, 59–60, 63, 70–1, 107, 117
Rugby football, 43–4
Rules, moral, 24–5, ch. 4 *passim*, 71–2, 83
Ryle, G., 66, 69, 147

SAKI (H. H. Munro), 61
Sanctions, 19
Sartre, J.-P., 35–6, 101

Savile, A., 64
Schlick, M., 120
Schoenberg, A. 99
Searle, J. R., 73
Sidgwick, H., 21, 109
Slater, O., 135
Smart, J. J. C., 50
Smith, Adam, 35, 101
Stevenson, C. L., 35, 79–91
Subjectivism (of duty), ch. 5 *passim*
Subjectivism (of moral values generally), ch. 9 *passim*
Synthetic *a priori* propositions, 92–4

TAYLOR, P., 64

UNITED Nations, 144
Universalisation, 39–49
Utilitarianism, chs. 2, 3 11 *passim*,

WELL-MEANING, 65–6
Wheatley, J., 64
Wimberley, H., 100
Wisdom, J. O., 98, 108

For Product Safety Concerns and Information please contact our EU representative GPSR@taylorandfrancis.com
Taylor & Francis Verlag GmbH, Kaufingerstraße 24, 80331 München, Germany

www.ingramcontent.com/pod-product-compliance
Lightning Source LLC
Chambersburg PA
CBHW052129300426
44116CB00010B/1825